THE OHIO STATE UNIVERSITY BULLETIN
VOLUME XVIII NUMBER 24

OHIO BIOLOGICAL SURVEY

BULLETIN 2

Catalog of

Ohio Vascular Plants

MARCH, 1914

PUBLISHED BY THE UNIVERSITY AT COLUMBUS, OHIO

Entered as second-class matter November 17, 1905, at the postoffice at Columbus, Ohio, under Act of Congress, July 16, 1894.

OHIO BIOLOGICAL SURVEY

HERBERT OSBORN, Director

OHIO STATE UNIVERSITY IN CO-OPERATION WITH OTHER OHIO COLLEGES AND UNIVERSITIES

ANNOUNCEMENT

The Bulletins of the Ohio Biological Survey will be issued as work on any special subject is completed, and will form volumes of about 500 pages each.

They will be sent to co-operating institutions and individuals, libraries and colleges in Ohio and to such surveys, societies and other organizations as may offer suitable exchange material.

Additional copies of each bulletin and of completed volumes will be sold at such price as may cover the cost of publication. Special rates on quantities to schools for classes.

Subscription for entire volumes, $2.00
Price of this number, .50

All orders should be accompanied by remittance which should be made payable to *Ohio Biological Survey* and sent to the Director.

Correspondence concerning the Survey, applications for exchanges and purchase of copies of Bulletins should be addressed to the Director—Professor Herbert Osborn, Columbus, Ohio.

VOLUME I BULLETIN 2

OHIO BIOLOGICAL SURVEY

CATALOG OF OHIO VASCULAR PLANTS

Arranged according to the phyletic classification; with notes on the geographical distribution in the state, based mainly on specimens in the State Herbarium, Botanical Laboratory, The Ohio State University.

BY

JOHN H. SCHAFFNER, M. A., M. S.

Published by
THE OHIO STATE UNIVERSITY
Columbus
1914

CS

CONTRIBUTION FROM THE BOTANICAL LABORATORY OF THE OHIO STATE UNIVERSITY, No. 83. : : : : :

PREFACE

This catalog of the vascular plants of Ohio is based on specimens for the most part in the State Herbarium of the Ohio State University, although some records have also been obtained from other collections. The State Herbarium is at present a collection of about 30,000 sheets, and represents the labors of many Ohio botanists. The distribution by counties is in many cases not the known distribution, but it was thought advisable to give only such data as could be verified by specimens. A list* of 254 species and varieties was published with a view to their exclusion from the state catalog if no evidence of their presence in the state were forthcoming. As a result of this publication, numerous records and specimens were obtained. The present catalog contains 2065 numbered species, about one-fourth of which are non-indigenous. Three additional species have been inserted in their proper places since the list was numbered for publication. A considerable number of varieties and supposed hybrids are also included.

"The Fourth State Catalogue of Ohio Plants," by W. A. Kellerman, 1899, was based on specimens in the State Herbarium at that time. However, a large number of species was included for which there was no direct evidence. Most of these, together with some species wrongly identified, have been omitted. A large list of contributors of specimens from various parts of the state was published in the "Fourth Catalogue." Many of these have greatly increased their collections in the State Herbarium and a number of other botanists have sent important specimens.

The species in the State Herbarium have been carefully determined, in the more difficult groups by the best experts in the country, and it is believed that there are few mistakes in the list as now published. Any errors, however, can be definitely corrected in the future. Several species became uncertain through the shifting of names and have been omitted until more and better material can be studied.

The species are numbered serially, and the varieties, forms, and supposed hybrids are indicated by letters. Additions of species will

*Plants on the Ohio State list not represented in the State Herbarium. Ohio Nat. 9: 413-415.

be made by means of the decimal system. For example, if a new species is to be added after No. 1866, it will be numbered 1866.1. Additions will only be made on the basis of good herbarium specimens. It is much better to have a small, reliable list than a large one with many doubtful entries. The introduced species have been designated by some phrase by which they can be distinguished from the indigenous species.

The nomenclature used is that of Britton and Brown's "An Illustrated Flora of the Northern United States, Canada, and the British Possessions," second edition. The few names which differ because of a different conception of the genus will cause no special trouble in reference. "Kellerman's Fourth State Catalogue of Ohio Plants" was based on the first edition of the "Illustrated Flora," and thus the names following the "American Code" on the principle of priority have been used very generally in Ohio since that time. The author sees no reason for abandoning the principle of priority at the present stage of progress of botanical knowledge.

The arrangement of the species and larger groups follows strictly the phyletic classification. The time has come when taxonomy must readjust itself to the more modern conceptions in regard to morphology and evolution.

The map of Ohio by counties will enable one to see, at a glance, the distribution indicated, and whether a given species is considered rare or unusual in any locality. It will now be possible to concentrate attention on the exact distribution of our more interesting plants and in the near future data should be at hand to definitely delimit the natural plant regions of the state. When this can be done, a considerable advance will have been made in the knowledge of the natural agricultural and horticultural regions of Ohio.

A number of useful lists have been published in the past which are still available. Among these may be mentioned the following:

Poisonous and Other Injurious Plants of Ohio. Ohio Nat. 4: 16-19; 32-35; 69-73. 1903-1904. By the author.

Medicinal Plants of Ohio. Ohio Nat. 10: 55-60; 73-85. 1910. By Freda Detmers.

The Non-Indigenous Flora of Ohio. Univ. Bull. Series 4, No. 27. 1900. By W. A. Kellerman and Mrs. Kellerman.

To the many botanists and collectors who have aided in the preparation of this catalog during the past five years, the author wishes to express his hearty thanks. It is hoped that all who take an interest in Ohio plants will continue to send collections and rare specimens to the State Herbarium. In this way alone can a truly great collection be accumulated.

J. H. SCHAFFNER,

Department of Botany, The Ohio State University.

January 1, 1914.

WILLIAMS
FULTON
LUCAS
OTTAWA
LAKE
ASHTABULA
GEAUGA
DEFIANCE
HENRY
WOOD
SANDUSKY
ERIE
LORAIN
CUYAHOGA
TRUMBULL
PORTAGE
SUMMIT
PAULDING
SENECA
HURON
MEDINA
PUTNAM
HANCOCK
MAHONING
VAN WERT
WYANDOT
CRAWFORD
RICHLAND
ASHLAND
WAYNE
STARK
ALLEN
COLUMBIANA
HARDIN
MERCER
AUGLAIZE
MARION
HOLMES
CARROLL
TUSKARAWAS
MORROW
JEFFERSON
LOGAN
SHELBY
UNION
DELAWARE
KNOX
COSHOCTON
HARRISON
CHAMPAIGN
LICKING
DARKE
MIAMI
FRANKLIN
MUSKINGUM
GUERNSEY
BELMONT
CLARK
MADISON
MONTGOMERY
PREBLE
GREENE
FAIRFIELD
NOBLE
PICKAWAY
FAYETTE
PERRY
MORGAN
MONROE
BUTLER
WARREN
CLINTON
HOCKING
WASHINGTON
ATHENS
ROSS
VINTON
HAMILTON
CLERMONT
HIGHLAND
BROWN
PIKE
MEIGS
JACKSON
GALLIA
ADAMS
SCIOTO
LAWRENCE
OHIO

Phylum, PTENOPHYTA

Class, FILICES. Ferns.

Subclass, EUSPORANGIATAE.

Order, *Ophioglossales.*

Ophioglossaceae. Adder-tongue Family.

1. Ophioglossum vulgatum L. Adder-tongue. Rather generally distributed but local.
2. Botrychium simplex Hitch. Little Grape-fern. Cedar Point, Erie County.
3. Botrychium neglectum Wood. Wood's Grape-fern. Northern counties.
4. Botrychium lanceolatum (Gmel.) Angs. Lanceleaf Grape-fern. Geauga, Portage.
5. Botrychium obliquum Muhl. Oblique Grape-fern. General.
6. Botrychium dissectum Spreng. Cutleaf Grape-fern. General.
7. Botrychium virginianum (L.) Sw. Virginia Grape-fern. General and common.

Subclass, LEPTOSPORANGIATAE.

Order, *Filicales.*

Osmundaceae. Royal-fern Family.

8. Osmunda regalis L. Royal-fern. General.
9. Osmunda claytoniana L. Clayton's Fern. General.
10. Osmunda cinnamomea L. Cinnamon-fern. General.

10a. Osmunda cinnamomea frondosa Gr. Wayne County.

Polypodiaceae. Polypody Family.

Subfamily, POLYPODIATAE.

11. Polypodium vulgare L. Common Polypody. General in the eastern half of the state.
12. Polypodium polypodioides (L.) Hitch. Gray Polypody. Adams, Hamilton.
13. Phegopteris phegopteris (L.) Und. (Dryopteris phegopteris (L.) Chr.) Long Beech-fern. Rather general but apparently local.

14. Phegopteris hexagonoptera (Mx.) Fee. (Dryopteris hexagonoptera (Mx.) Chr). Broad Beech-fern. General.
15. Phegopteris dryopteris (L.) Fee. (Dryopteris dryopteris (L.) Britt). Oak-fern. Geauga, Lake, Ashtabula, Wayne, Summit.

Subfamily, PTERIDATAE.

16. Adiantum pedatum L. Maiden-hair-fern. General and common.
16a. Adiantum pedatum laciniatum Hopkins. Wayne County.
17. Pteridium aquilinum (L.) Kuhn. Eagle-fern. General.
18. Pellaea atropurpurea (L.) Link. Purple Cliff-brake. Ottawa, Stark, Franklin, Licking, Clark, Greene, Highland, Adams.

Subfamily, ASPLENIATAE.

19. Anchistea virginica (L.) Presl. Virginia Chain-fern. Ashtabula, Defiance, Geauga, Portage, Wayne, Williams.
20. Asplenium pinnatifidum Nutt. Pinnatifid Spleenwort. Mahoning, Licking, Fairfield, Hocking, Lawrence.
21. Asplenium platyneuron (L.) Oakes. Ebony Spleenwort. General except in the northeastern fourth of the state.
21a. Asplenium platyneuron x Camptosorus rhizophyllus (Asplenium ebenoides Scott). Hocking County.
22. Asplenium resiliens Kunze. Small Spleenwort. Adams County.
23. Asplenium trichomanes L. Maidenhair Spleenwort. General except in the northwestern fourth of the state.
24. Asplenium pycnocarpon Spreng. Narrow-leaf Spleenwort. General.
25. Asplenium ruta-muraria L. Wall-rue Spleenwort. Highland, Greene.
26. Asplenium montanum Willd. Mountain Spleenwort. Hocking, Fairfield, Summit, Mahoning, (Tuscarawas—Hopkins.)
27. Athyrium thelypteroides (Mx.) Desv. Silvery Spleenwort. General.
28. Athyrium filix-foemina (L.) Roth. Lady-fern. General.
29. Camptosorus rhizophyllus (L.) Link. Walking-fern. General.

Subfamily, DRYOPTERIDATAE.

30. Dryopteris noveboracensis (L.) Gr. New York Shield-fern. General.

31. Dryopteris thelypteris (L.) Gr. Marsh Shield-fern. General.
32. Dryopteris cristata (L.) Gr. Crested Shield-fern. General.
32a. Dryopteris cristata x spinulosa. Wayne, Portage.
32b. Dryopteris cristata x intermedia. (D. boottii (Tuck.) Und.). Wayne, Geauga, Portage.
33. Dryopteris clintoniana (Eat.) Dow. Wayne, Geauga.
33a. Dryopteris clintoniana X spinulosa. Wayne County.
34. Dryopteris goldiana (Hook.) Gr. Goldie's Shield-fern. Rather general, but no specimens from the southern third of the state, nor from the northwestern counties.
35 Dryopteris intermedia (Muhl.) Gr. American Shield-fern. Rather general.
35a. Dryopteris intermedia x marginalis. Wayne County—Hopkins.
36. Dryopteris spinulosa (Muell.) Ktz. Spinulose Shield-fern. General.
37. Dryopteris dilatata (Hoffm.) Gr. Spreading Shield-fern. Tuscarawas County.
38. Dryopteris marginalis (L.) Gr. Marginal Shield-fern. General and common.
39 Polystichum acrostichoides (Mx.) Schott. Christmas-fern. General.
39a. Polystichum acrostichoides schweinitzii (Beck) Small. Wayne County.
40. Dennstaedtia punctilobula (Mx.) Moore. Boulder-fern. In the southern and eastern parts of the state. (Erie County—Moseley Herb.)
41. Filix bulbifera (L.) Und. Bulbiferous Bladder-fern. (Cystopteris.) General, but no specimens from the northwest.
42. Filix fragilis (L.) Und. Fragile Bladder-fern. General and common.
42a. Filix fragilis magnasora (Clute). Tuscarawas County.
42b. Filix fragilis cristata (Hopkins). Portage County.
43. Woodsia obtusa (Spreng.) Torr. Blunt-lobed Woodsia. In the southern half of the state.
44. Matteuccia struthiopteris (L.) Todaro. Ostrich-fern. Cuyahoga, (Erie County—Moseley Herb.).
45. Onoclea sensibilis L. Sensitive-fern. General and common.

Class, HYDROPTERIDAE. Water-ferns.

Order, *Marsileales.*

Marsileaceae. Marsilea Family.

46. Marsilea quadrifolia L. European Marsilea. A waif in Franklin County.

Order, *Salviniales.*

Salviniaceae. Salvinia Family.

47. Azolla caroliniana Willd. Carolina Azolla. Hamilton, Lucas, Lake.

Class, ISOETEAE. Quillworts.

Order, *Isoetales.*

Isoetaceae. Quillwort Family.

48. Isoetes braunii Durieu. Braun's Quillwort. Lake Brady, Portage County—Hopkins.
49. Isoetes foveolata Eat. Pitted Quillwort. Lake Brady, Portage County—Hopkins.

Phylum, CALAMOPHYTA

Class, EQUISETEAE. Horsetails and Scouring-rushes.

Order, *Equisetales.*

Equisetaceae. Horsetail Family.

50. Equisetum hyemale L. Common Scouring-rush. General.
51. Equisetum prealtum Raf. Great Scouring-rush. (E. robustum A. Br.). General in the state.
52. Equisetum variegatum Schleich. Variegated Scouring-rush. Lake, Erie.
53. Equisetum laevigatum A. Br. Smooth Scouring-rush. General but apparently not common.
54. Equisetum fluviatile L. Swamp Horsetail. Not common but to be found in most parts of the state in suitable places.
55. Equisetum sylvaticum L. Wood Horsetail. Auglaize, Cuyahoga, Geauga.
56. Equisetum pratense Ehrh. Thicket Horsetail. Supposed to occur in the state but no definite specimens.
57. Equisetum arvense L. Field Horsetail. General and abundant.

Phylum, LEPIDOPHYTA

Class, LYCOPODIEAE. Lycopods.

Order, *Lycopodiales.*

Lycopodiaceae. Club-moss Family.

58. Lycopodium lucidulum Mx. Shining Club-moss. General in the eastern half of the state.
59. Lycopodium porophilum Lloyd & Und. Rock Club-moss. Licking, Fairfield, Hocking, Portage.
60. Lycopodium inundatum L. Bog Club-moss. Portage County.
61. Lycopodium clavatum L. Common Club-moss. Ashtabula, Cuyahoga, Geauga, Portage, Hocking, (Stark—Hopkins).
62. Lycopodium obscurum L. Tree Club-moss. Ashtabula, Lake, Medina, Portage, Licking, Defiance, Fairfield, Hocking.
63. Lycopodium complanatum L. Trailing Club-moss. Ashtabula, Cuyahoga, Carroll, Geauga, Licking, Fairfield, Hocking, Portage, Lake, (Erie—Moseley), (Wayne—Hopkins).

Class, SELAGINELLEAE. Selaginellas.

Order, *Selaginellales.*

Selaginellaceae. Selaginella Family.

64. Selaginella rupestris (L.) Spring. Rock Selaginella. Licking, Fairfield, Hocking.
65. Selaginella apus (L.) Spring. Creeping Selaginella. Lake, Trumbull.

Phylum, STROBILOPHYTA

Class, CONIFERAE. Conifers.

Order, *Pinales.*

Pinaceae. Pine Family.

66. Tsuga canadensis (L.) Carr. Hemlock. Eastern half of Ohio; occasional toward the west.
67. Larix laricina (DuR.) Koch. Tamarack. Northern third of the state.
68. Pinus strobus L. White Pine. Northern part of Ohio.
69. Pinus rigida Mill. Pitch Pine. Lawrence, Scioto, Jackson, Fairfield.

70. Pinus virginiana Mill. Scrub Pine. From Licking County southward.
71. Pinus echinata Mill. Yellow Pine. Auglaize County. Probably accidental from seed from cultivated trees. Not native.

Juniperaceae. Juniper Family.

Subfamily, CUPRESSATAE.

72. Thuja occidentalis L. Arborvitae. Champaign, Franklin, Greene, Highland, Adams.

Subfamily, JUNIPERATAE.

73. Juniperus communis L. Common Juniper. Northern Ohio, as far south as Fairfield County.
74. Juniperus sibirica Burgs. Low Juniper. Erie County.
75. Juniperus virginiana L. Red Juniper. General.

Order, *Taxales.*

Taxaceae. Yew Family.

76. Taxus canadensis Marsh. American Yew. Northern Ohio, as far south as Hocking, Highland, and Greene Counties.

Phylum, ANTHOPHYTA

Class, MONOCOTYLAE. Monocotyls.

Subclass, HELOBIAE.

Order, *Alismales.*

Alismaceae. Water-plantain Family.

77. Lophotocarpus calycinus (Eng.) Sm. Large Lophotocarpus. Erie, Ottawa, Auglaize.
78. Sagittaria latifolia Willd. Broadleaf Arrow-head. General and abundant.
79. Sagittaria cuneata Sheld. Arum-leaf Arrow-head. (Erie County—Moseley Herbarium.)
80. Sagittaria rigida Pursh. Sessile-fruited Arrow-head. Mostly in the northern counties.
81. Sagittaria graminea Mx. Grassleaf Arrow-head. Lucas County.
82. Alisma subcordatum Raf. American Water-plantain. General.

Scheuchzeriaceae. Arrow-grass Family.

83. Triglochin palustris L. Marsh Arrow-grass. Erie, Madison.
84. Triglochin maritima L. Seaside Arrow-grass. Summit, Stark, Clark, Champaign.
85. Scheuchzeria palustris L. Scheuchzeria. Licking, Ashtabula.

Potamogetonaceae. Pondweed Family.

86. Potamogeton natans L. Common Floating Pondweed. General.
87. Potamogeton amplifolius Tuck. Large-leaf Pondweed. Wayne, Stark, Summit, Erie.
88. Potamogeton epihydrus Raf. Nuttall's Pondweed. Trumbull, Wayne.
89. Potamogeton americanus Cham. & Schl. Longleaf Pondweed. Rather general.
90. Potamogeton heterophyllus Schreb. Variant-leaf Pondweed. Ottawa, Erie, Wayne, Stark, Ashtabula.
91. Potamogeton angustifolius Berch & Presl. Narrowleaf Pondweed. Wayne County.
92. Potamogeton lucens L. Shining Pondweed. Summit, Cuyahoga, Wayne, Erie.
93. Potamogeton praelongus Wulf. White-stem Pondweed. Ashtabula, Wayne.
94. Potamogeton perfoliatus L. Clasping-leaf Pondweed. Erie, Summit.
95. Potamogeton compressus L. Eel-grass Pondweed. Perry, Logan, Erie, Stark.
96. Potamogeton hillii Mor. Hill's Pondweed. (Ottawa County—Moseley Herbarium.)
97. Potamogeton foliosus Raf. Leafy Pondweed. General.
98. Potamogeton obtusifolius Mert. & Koch. Bluntleaf Pondweed. Medina County.
99. Potamogeton friesii Rupr. Fries' Pondweed. (Erie County—Moseley Herbarium.)
100. Potamogeton pusillus L. Small Pondweed. Erie, Auglaize, Fairfield, Perry, Summit.
101. Potamogeton diversifolius Raf. Rafinesque's Pondweed. Ashtabula, Portage.

102. Potamogeton pectinatus L. Fennel-leaf Pondweed. General.
103. Potamogeton interruptus Kit. Interrupted Pondweed. Erie County.
104. Potamogeton robbinsii Oakes. Robbins' Pondweed. Erie, Summit.
105. Zannichellia palustris L. Zannachellia. General.

Naiadaceae. Naias Family.

106. Naias flexilis (Willd.) Rost. & Schm. Slender Naias. General.
107. Naias gracillima (A. Br.) Magnus. Thread-like Naias. Wayne County.

Order, *Nymphaeales.*

Nymphaeaceae. Water-lily Family.

Subfamily, CABOMBATAE.

108. Brasenia schreberi Gmel. Water-shield. Geauga, Summit, Portage, Stark, Wayne.

Subfamily, NELUMBONATAE.

109. Nelumbo lutea (Willd.) Pers. American Water-lotus. Licking, Perry, Erie, Auglaize.

Subfamily, NYMPHAEATAE.

110. Nymphaea advena Sol. Large yellow Water-lily. General.
111. Castalia odorata (Dry.) W. & W. Sweet-scented White Water-lily. Rather general.
112. Castalia tuberosa (Paine) Greene. Tuberous White Water-lily. Erie, Holmes, Licking.

Order, *Hydrocharitales.*

Vallisneriaceae. Tape-grass Family.

113. Philotria canadensis (Mx.) Britt. Common Water-weed. General.
114. Philotria minor (Eng.) Small. Lesser Water-weed. No specimens.
115. Vallisneria spiralis L. Tape-grass. Shelby, Fairfield, Stark, Summit, Geauga, Erie.

Subclass, SPADICIFLORAE.

Order, *Pandanales.*

Sparganiaceae. Bur-reed Family.

116. Sparganium eurycarpum Eng. Broad-fruited Bur-reed. General.
117. Sparganium androcladum (Eng.) Morong. Branching Bur-reed. Franklin, Auglaize, Lucas.
118. Sparganium lucidum Fern. & Eames. Shining-fruited Bur-reed. Richland County.
119. Sparganium simplex Huds. Simple-stemmed Bur-reed. (Erie County—Moseley Herbarium.)

Typhaceae. Cat-tail Family.

120. Typha latifolia L. Broad-leaf Cat-tail. General.
121. Typha angustifolia L. Narrow-leaf Cat-tail. Auglaize, Licking, Knox, Erie, Cuyahoga, Geauga.

Order, *Arales.*

Araceae. Arum Family.

Subfamily, POTHATAE.

122. Acorus calamus L. Sweet-flag. General.

Subfamily, CALLATAE.

123. Calla palustris L. Wild Calla. Ashtabula, Portage, Summit, Stark.
124. Spathyema foetida (L.) Raf. Skunk-cabbage. General.

Subfamily, PHILODENDRATAE.

125. Peltrandra virginica (L.) Kunth. Green Arrow-arum. Cuyahoga, Summit, Stark, Licking, Perry.

Subfamily, ARATAE.

126. Arisaema triphyllum (L.) Torr. Jack-in-the-pulpit. General.
127. Arisaema dracontium (L.) Schott. Green-dragon. General.

Lemnaceae. Duckweed Family.

128. Spirodela polyrhiza (L.) Schl. Greater Duckweed. General.

129. Lemna trisulca L. Ivy-jointed Duckweed. General.
130. Lemna cyclostasa (Ell.) Chev. Valdivia Duckweed. Lake County.
131. Lemna minor L. Lesser Duckweed. General.
132. Wolffiella floridiana (J. D. Sm.) Thomp. Florida Wolffiella. Licking County.
133. Wolffia columbiana Karst. Columbia Wolffia. Ottawa, Franklin.
134. Wolffia punctata Griseb. Punctate Wolffia. Erie County.

Subclass, GLUMIFLORAE.

Order, *Graminales.*

Cyperaceae. Sedge Family.

Subfamily, SCIRPATAE.

135. Cyperus schweinitzii Torr. Schweinitz's Cyperus. Erie, Cuyahoga.
136. Cyperus esculentus L. Nut-grass (Cyperus). Western half of state, as far east as Wayne County.
137. Cyperus erythrorhizos Muhl. Red-rooted Cyperus. Rather general.
138. Cyperus inflexus Muhl. Awned Cyperus. Lucas, Champaign.
139. Cyperus strigosus L. Straw-colored Cyperus. General.
140. Cyperus filiculmis Vahl. Slender Cyperus. Rather general.
141. Cyperus engelmanni Steud. Engelmann's Cyperus. Wayne, Logan.
142. Cyperus speciosus Vahl. Michaux's Cyperus. Ashtabula, Erie.
143. Cyperus flavescens L. Yellow Cyperus. Fairfield, Richland, Meigs.
144. Cyperus diandrus Torr. Low Cyperus. Northern half of state, as far south as Champaign County.
145. Cyperus rivularis Kunth. Shining Cyperus. Rather general.
146. Kyllinga pumila Mx. Low Kyllinga. Fairfield, Hocking, Cuyahoga, Auglaize.
147. Dulichium arundinaceum (L.) Britt. Dulichium. Northern part of the state, south to Hocking and Clark Counties.
148. Eleocharis mutata (L.) R. & S. Four-angled Spike-rush. Ashland County.

149. Eleocharis olivacea Torr. Olivaceous Spike-rush. Auglaize, Licking, Summit, Cuyahoga.
150. Eleocharis ovata (Roth) R. & S. Ovoid Spike-rush. Erie County.
151. Eleocharis obtusa (Willd.) Schul. Blunt Spike-rush. General.
152. Eleocharis engelmanni Steud. Engelmann's Spike-rush. Licking County.
153. Eleocharis palustris (L.) R. & S. Creeping Spike-rush. General.
154. Eleocharis acicularis (L.) R. & S. Needle Spike-rush. General.
155. Eleocharis tenuis (Willd.) Schultes. Slender Spike-rush. Lucas, Ottawa.
156. Eleocharis acuminata (Muhl.) Nees. Flat-stemmed Spike-rush. Lucas, Ottawa, Auglaize.
157. Eleocharis intermedia (Muhl.) Schultes. Matted Spike-rush. Erie, Franklin.
158. Stenophyllus capillaris (L.) Britt. Hair-like Stenophyllus. Lucas County.
159. Fimbristylis autumnalis (L.) R. & S. Slender Fimbristylis. Hamilton, Defiance, Fairfield, Hocking.
160. Scirpus cyperinus (L.) Kunth. Wool-grass. General.
161. Scirpus lineatus Mx. Reddish Bulrush. General.
162. Scirpus polyphyllus Vahl. Leafy Bulrush. Eastern half of the state to Crawford and Adams Counties.
163. Scirpus atrovirens Muhl. Dark-green Bulrush. General.
164. Scirpus sylvaticus L. Wood Bulrush. Ottawa, Wayne, Hamilton.
165. Scirpus fluviatilis (Torr.) Gr. River Bulrush. Champaign, Licking, Lucas, Wayne, Medina, Lake.
166. Scirpus validus Vahl. Great Bulrush. General.
167. Scirpus torreyi Olney. Torrey's Club-rush. Lake, Erie.
168. Scirpus americanus Pers. Chair-maker's Club-rush. General.
169. Scirpus debilis Pursh. Weak Club-rush. Summit, Ottawa.
170. Scirpus planifolius Muhl. Flat-leaf Club-rush. Licking, Knox, Lake.
171 Eriophorum viridicarinatum (Eng.) Fern. Thin-leaf Cotton-grass. Geauga, Summit, Licking.
172. Eriophorum virginicum L. Virginia Cotton-grass. Northern Ohio, as far south as Licking County.

Subfamily, RYNCHOSPORATAE.

173. Rynchospora corniculata (Lam.) Gr. Horned-rush. None in the herbarium.
174. Rynchospora alba (L.) Vahl. White Beaked-rush. Geauga, Lorain, Summit, Ashland, Stark, Licking, Champaign.
175. Rynchospora capillacea Torr. Capillary Beaked-rush. Madison, Greene, Champaign, Erie.
176. Rynchospora glomerata (L.) Vahl. Clustered Beak-rush. Erie, Ashtabula, Portage, Summit, Fairfield.
177. Rynchospora cymosa Ell. Grass-like Beak-rush. Erie County.
178. Mariscus mariscoides (Muhl.) Kuntze. Twig-rush. Erie County.
179. Scleria triglomerata Mx. Tall Nut-rush. Erie County.
180. Scleria pauciflora Muhl. Papillose Nut-rush. Erie County.
181. Scleria verticillata Muhl. Low Nut-rush. Erie, Franklin, Champaign.

Subfamily, CARICATAE.

182. Carex sartwellii Dew. Sartwell's Sedge. Erie County.
183. Carex siccata Dew. Dry-spiked Sedge. No specimens. (Erie County—Moseley Herbarium.)
184. Carex retroflexa Muhl. Reflexed Sedge. From Erie, Huron, Wayne, Knox and Perry Counties westward.
185. Carex rosea Schk. Stellate Sedge. General.
186. Carex muricata L. Lesser Prickly Sedge. No specimens. (Erie County—Moseley Herbarium.) Naturalized from Europe.
187. Carex muhlenbergii Schk. Muhlenberg's Sedge. Carroll, Summit, Cuyahoga, Erie, Lucas.
188. Carex cephalophora Muhl. Oval-headed Sedge. General.
189. Carex gravida Bail. Heavy Sedge. No specimens.
190. Carex cephaloidea Dew. Thinleaf Sedge. Cuyahoga, Licking.
191. Carex sparganioides Muhl. Bur-reed Sedge. General.
192. Carex conjuncta Boott. Soft Fox Sedge. General, but no specimens from the eastern and southern counties.
193. Carex vulpinoidea Mx. Fox Sedge. General and abundant.
194. Carex setacea Dew. Bristly-spiked Sedge. Erie County.
195. Carex diandra Schr. Lesser Panicled Sedge. Lake County.
196. Carex prairea Dew. Prairie Sedge. Summit, Erie.
197. Carex decomposita Muhl. Large-panicled Sedge. Licking County.

198. Carex stipitata Muhl. Awl-fruited Sedge. General, but no specimens from the extreme southern and southeastern counties.
199. Carex crus-corvi Shuttlw. Raven-foot Sedge. Defiance, Auglaize, Wayne.
200. Carex disperma Dew. Soft-leaf Sedge. Erie County.
201. Carex trisperma Dew. Three-fruited Sedge. Lake, Portage, Summit, Williams.
202. Carex canescens L. Silvery Sedge. Summit, Lorain, Logan.
203. Carex brunnescens (Pers.) Poir. Brownish Sedge. Logan County.
204. Carex deweyana Schw. Dewey's Sedge. Auglaize County.
205. Carex bromoides Schk. Brome-like Sedge. Cuyahoga, Erie, Hancock, Hardin, Auglaize.
206. Carex interior Bail. Inland Sedge. Cuyahoga, Summit, Erie, Stark, Licking.
207. Carex leersii Willd. Little Prickly Sedge. Ottawa, Erie, Madison, Geauga, Portage.
208. Carex scoparia Schk. Pointed Broom Sedge. Northern Ohio, as far south as Auglaize, Madison and Tuscarawas Counties.
209. Carex tribuloides Wahl. Blunt Broom Sedge. General.
210. Carex cristatella Britt. Crested Sedge. General.
211. Carex muskingumensis Schw. Muskingum Sedge. Defiance, Auglaize, Wyandot, Franklin, Champaign.
212. Carex bebbii Olney. Bebb's Sedge. Franklin County.
213. Carex straminea Willd. Straw Sedge. Cuyahoga, Portage, Lucas, Williams.
214. Carex normalis Mack. Larger Straw Sedge. Hamilton, Auglaize, Wayne, Hancock, (Erie County—Moseley Herbarium).
215. Carex festucacea Schk. Fescue Sedge. Lake, Erie, Stark, Logan, Madison.
216. Carex bicknellii Britt. Bicknell's Sedge. Lake County, (Erie County—Moseley Herbarium).
217. Carex alata Torr. Broad-winged Sedge. Summit County.
218. Carex albolutescens Schw. Greenish-white Sedge. Champaign County.
219. Carex foenea Willd. Hay Sedge. Geauga County.
220. Carex willdenovii Schk. Willdenow's Sedge. Cuyahoga County.
221. Carex jamesii Schw. James' Sedge. General.
222. Carex durifolia Bail. Back's Sedge. No specimens.

223. Carex leptalea Wahl. Bristle-stalked Sedge. Lake, Cuyahoga, Summit, Madison.

224. Carex communis Bail. Fibrous-rooted Sedge. Summit County. (Erie County—Moseley Herbarium).

225. Carex pennsylvanica Lam. Pennsylvania Sedge. General.

226. Carex varia Muhl. Emmons' Sedge. From Erie, Madison, and Clermont Counties eastward.

227. Carex hirtifolia Mack. Pubescent Sedge. From Lake, Wayne, Delaware, and Greene Counties northwestward.

228. Carex pedunculata Muhl. Long-stalked Sedge. Lake, Cuyahoga, Geauga, Erie.

229. Carex richardsonii R. Br. Richardson's Sedge. (Erie County —Moseley Herbarium.)

230. Carex eburnea Boott. Bristle-leaf Sedge. Greene, Ottawa.

231. Carex aurea Nutt. Golden-fruited Sedge. Erie County.

232. Carex meadii Dew. Mead's Sedge. Erie County.

233. Carex tetanica Schk. Wood's Sedge. Geauga, Cuyahoga, Erie, Huron, Auglaize.

234. Carex plantaginea Lam. Plantain-leaf Sedge. Fairfield, Delaware, Huron, Lorain, Cuyahoga, Summit.

235. Carex careyana Torr. Carey's Sedge. Lorain County.

236. Carex platyphylla Car. Broadleaf Sedge. From Cuyahoga, Knox, Fairfield, and Hocking Counties eastward.

237. Carex digitalis Willd. Slender Wood Sedge. Rather general.

238. Carex laxiculmis Schw. Spreading Sedge. Lake, Cuyahoga, Auglaize, Franklin.

239. Carex albursina Sheld. White Bear Sedge. General.

240. Carex blanda Dew. Woodland Sedge. Rather general.

241. Calex laxiflora Lam. Loose-flowered Sedge. Rather general.

242. Carex anceps Muhl. Two-edged Sedge. Rather general.

243. Carex shriveri Britt. Shriver's Sedge. No specimens.

244. Carex granularis Muhl. Meadow Sedge. General, but no specimens from the southeastern counties.

245. Carex crawei Dew. Crawe's Sedge. Erie County.

246. Carex oligocarpa Schk. Few-fruited Sedge. Hamilton, Montgomery, Greene, Ross.

247. Carex hitchcockiana Dew. Hitchcock's Sedge. Butler, Highland, Auglaize.

248. Carex conoidea Schk. Field Sedge. Wood County.
249. Carex amphibola Steud. Narrow-leaf Sedge. Cuyahoga, Hardin, Auglaize.
250. Carex grisea Wahl. Gray Sedge. Rather general, but the only southern county represented is Lawrence.
251. Carex glaucodea Tuck. Glaucescent Sedge. Cuyahoga County.
252. Carex gracillima Schw. Graceful Sedge. Northern Ohio, as far south as Licking and Franklin Counties.
253. Carex prasina Wahl. Drooping Sedge. Trumbull, Cuyahoga, Crawford, Fairfield, Hamilton.
254. Carex davisii Schw. Davis' Sedge. Lake, Lorain, Wayne, Auglaize, Licking, Ross.
255. Carex flexuosa Muhl. Slender-stalked Sedge. Cuyahoga, Geauga, Lake.
256. Carex arctata Boott. Drooping Wood Sedge. Summit County, (Erie County—Moseley Herbarium).
257. Carex virescens Muhl. Ribbed Sedge. Northern Ohio, as far south as Geauga, Hocking, and Auglaize Counties.
258. Carex complanata Torr. Hirsute Sedge. General.
259. Carex scabrata Schw. Rough Sedge. Lake, Cuyahoga, Summit.
260. Carex limosa L. Mud Sedge. No specimens.
261. Carex paupercula Mx. Bog Sedge. Licking County.
262. Carex buxbaumii Wahl. Brown Sedge. Lorain, Erie, Lucas.
263. Carex shortiana Dew. Short's Sedge. General; no specimens from the southeastern counties.
264. Carex stricta Lam. Tussock Sedge. From Trumbull, Madison, Champaign, and Auglaize Counties, northward.
265. Carex haydeni Dew. Hayden's Sedge. No specimens; (Moseley Herbarium—Erie County).
266. Carex torta Boott. Twisted Sedge. Cuyahoga, Erie, Hardin, Knox, Delaware, Franklin.
267. Carex aquatilis Wahl. Water Sedge. Auglaize, Lucas.
268. Carex gynandra Schw. Nodding Sedge. Harrison, Perry.
269. Carex crinita Lam. Fringed Sedge. General.
270. Carex lacustris Willd. Lake-bank Sedge. Northern Ohio, south to Summit, Morrow, and Logan Counties.
271. Carex impressa (Wright) Mack. Wright's Sedge. No specimens.
272. Carex lanuginosa Mx. Woolly Sedge. From Ashtabula, Franklin, Madison, Champaign, and Auglaize Counties northward.

273. Carex lasiocarpa Ehrh. Slender Sedge. Licking, Stark.
274. Carex trichocarpa Muhl. Hairy-fruited Sedge. Lorain, Erie, Licking, Madison.
275. Carex atherodes Spreng. Awned Sedge. Erie County.
276. Carex oederi Retz. Green Sedge. Erie County
277. Carex flava L. Yellow Sedge. Lake County.
278. Carex folliculata L. Long Sedge. Crawford, Portage.
279. Carex monile Tuck. Necklace Sedge. Lake, Cuyahoga, Wayne, Lucas.
280. Carex vesicaria L. Inflated Sedge. No specimens.
281. Carex rostrata Stokes. Beaked Sedge. Licking, Geauga.
282. Carex tuckermanii Dew. Tuckerman's Sedge. Northern Ohio, south to Auglaize, Franklin, and Perry Counties.
283. Carex retrorsa Schw. Retrorse Sedge. Lucas County
284. Carex oligosperma Mx. Few-seeded Sedge. Defiance County.
285. Carex lurida Wahl. Sallow Sedge. General; no specimens from the northwestern counties.
286. Carex hystricina Muhl. Porcupine Sedge. Rather general.
287. Carex pseudo-cyperus L. Cyperus-like Sedge. Greene, Tuscarawas, Wayne.
288. Carex comosa Boott. Bristly Sedge. Northern part of the state, as far south as Champaign, Franklin, and Tuscarawas Counties.
289. Carex frankii Kunth. Frank's Sedge. General as far north as Auglaize, Knox, and Jefferson Counties; also in Wyandot, Erie, and Cuyahoga.
290. Carex squarrosa L. Squarrose Sedge. General.
291. Carex typhina Mx. Cat-tail Sedge. Lake, Cuyahoga, Morrow, Scioto.
292. Carex intumescens Rudge. Bladder Sedge. Northern Ohio, as far south as Trumbull, Crawford, and Allen Counties.
293. Carex asa-grayi Bail. Gray's Sedge. General.
294. Carex lupulina Muhl. Hop Sedge. Rather general.
295. Carex lupuliformis Sartw. Hop-like Sedge. Cuyahoga, Wayne, Marion.

Graminaceae. Grass Family

Subfamily, POACATAE.

296. Bromus brizaeformis Fisch. & Mey Awnless Chess. Introduced. Cuyahoga County.

297. Bromus kalmii Gr. Kalm's Chess. Franklin, Lucas.
298. Bromus hordeaceus L. Soft Chess. Introduced. Wayne, Lorain.
299 Bromus secalinus L. Common Chess. Naturalized. General and abundant.
300. Bromus racemosus L. Upright Chess. Naturalized. General and abundant.
301. Bromus arvensis L. Field Chess. Introduced. Franklin County.
302. Bromus inermis Leyss. Hungarian Brome-grass. Introduced. Wayne County.
303. Bromus ciliatus L. Fringed Brome-grass. Erie, Wayne, Franklin, Champaign, Hocking.
304. Bromus purgans L. Hairy Brome-grass. General.
305. Bromus asper Murr. Rough Brome-grass. No specimens. European.
306. Bromus tectorum L. Downy Brome-grass. Introduced. General and abundant.
307. Bromus sterilis L. Barren Brome-grass. Introduced. Licking, Sandusky, Cuyahoga.
308. Melica nitens Nutt. Tall Melic-grass. Erie County.
309. Festuca elatior L. Tall Fescue-grass. Introduced. General.
310. Festuca nutans Willd. Nodding Fescue-grass. General.
311. Festuca ovina L. Sheep Fescue-grass. From Europe. Erie, Franklin, Wayne.
312. Festuca capillata Lam. Filiform Fescue-grass. From Europe. Cuyahoga County.
313. Festuca octoflora Walt. Slender Fescue-grass. Ashtabula, Erie, Lucas, Ashland, Delaware, Licking, Lawrence
314. Festuca myuros L. Rat-tail Fescue-gress. From Europe. Lake County.
315. Panicularia acutiflora (Torr) Ktz. Sharp-glumed Manna-grass. No specimens.
316. Panicularia fluitans (L.) Ktz. Floating Manna-grass. General as far south as Harrison, Perry, Franklin, and Auglaize Counties.
317. Panicularia canadensis (Mx.) Ktz. Rattlesnake Manna-grass. Ashtabula, Geauga, Portage, Cuyahoga, Lorain, Summit, Stark, Wayne.

318. Panicularia torreyana (Spreng.) Merr. Long Manna-grass. Ashtabula, Cuyahoga, Summit, Erie, Fairfield.

319. Panicularia nervata (Willd.) Ktz. Nerved Manna-grass. General.

320. Panicularia grandis (Wats.) Nash. Tall Manna-grass. Stark, Wayne.

321. Panicularia pallida (Torr.) Ktz. Pale Manna-grass. (Ottawa County—Moseley Herbarium.)

322. Poa compressa L. Flat-stemmed Blue-grass. From Europe. General and abundant.

323. Poa trivialis L. Rough-stalked Meadow-grass. From Europe. Crawford County.

324. Poa debilis Torr. Weak Spear-grass. (Erie County—Moseley Herbarium.)

325. Poa triflora Gilib. Fowl Meadow-grass. Fairfield, Geauga, Lawrence.

326. Poa nemoralis L. Wood Meadow-grass. Introduced. Lake County.

327. Poa pratensis L. Kentucky Blue-grass. General and abundant.

328. Poa autumnalis Muhl. Flexuous Spear-grass. Hocking County.

329. Poa sylvestris Gr. Sylvan Spear-grass. Rather general; no specimens from the northwestern counties.

330. Poa alsodes Gr. Grove Meadow-grass. Seneca, Franklin, Summit, Cuyahoga, Trumbull, Knox.

331. Poa brachyphylla Schult. Short-leaf Spear-grass. Lawrence, Perry, Medina, Cuyahoga, Trumbull.

332. Poa annua L. Annual Meadow-grass. From Europe. General.

333. Dactylis glomerata L. Orchard-grass. Naturalized. General; no specimens from the southeastern counties.

334. Eragrostis pectinacea (Mx.) Steud. Purple Love-grass. Lake, Cuyahoga, Erie, Auglaize.

335. Eragrostis hypnoides (Lam.) B. S. P. Creeping Love-grass. Rather general; no specimens from the central eastern counties.

336. Eragrostis major Host. Strong-scented Love-grass. Naturalized. General.

337. Eragrostis purshii Schrad. Pursh's Love-grass. Erie County.

338. Eragrostis pilosa (L.) Beauv. Tufted Love-grass. Rather general; no specimens from the northwestern counties. Naturalized.

339. Eragrostis frankii Steud. Frank's Love-grass. Rather general; no specimens from the southeastern nor from the northwestern counties.

340. Eragrostis capillaris (L.) Nees. Capillary Love-grass. Ottawa, Madison, Clinton.

341. Sphenopholis obtusata (Mx.) Scribn. Blunt-scaled Eaton-grass. No specimens.

342. Sphenopholis pallens (Spreng.) Scribn. Tall Eaton-grass. General.

343. Sphenopholis nitida (Spreng.) Scribn. Slender Eaton-grass. Cuyahoga, Knox, Licking, Fairfield, Hocking, Lawrence, Adams.

344 Koeleria cristata (L.) Pers. Crested Koeler-grass. Ottawa County.

345. Korycarpus arundinaceus Zea. American Korycarpus. Ross, Franklin, Auglaize, Highland.

346. Tridens flava (L.) Hitch. Tall Purple-top. Rather general; no specimens from the northwestern counties nor the extreme eastern part.

347. Triplasis purpurea (Walt.) Chapm. Purple Sand-grass. Ashtabula, Cuyahoga, Erie.

348. Cynosurus cristatus L. Dogtail-grass. From Europe. Mahoning County.

349. Phragmites phragmites (L.) Karst. Common Reed-grass. Ashtaubula, Cuyahog, Erie, Lucas, Huron, Wayne, Franklin.

350. Danthonia spicata (L.) Beauv. Common Wild-oat-grass. General.

351. Danthonia compressa Aust. Flattened Wild-oat-grass. Portage County.

352. Arrenatherum elatius (L.) Beauv. Oat-grass. From Europe. Hamilton County.

353. Trisetum pennsylvanicum (L.) Beauv. Marsh False-oats. No specimens.

354. Avena torreyi Nash. Purple Oats. Franklin County.

355. Avena sativa L. Common Oats. Rather general. Escaped from cultivation.

356. Avena fatua L. Wild Oats. From Europe. No specimens.

357. Deschampsia flexuosa (L.) Trin. Wavy Hair-grass. Portage County.

358. Aspris caryophyllea (L.) Nash. Silvery Hair-grass. From Europe. Lake County.

359. Nothoholcus lanatus (L.) Nash. Velvet-grass. Lake, Trumbull, Cuyahoga, Lorain, Erie, Wayne, Fairfield. From Europe.

360. Lolium perenne L. Red Darnel. Rather general. From Europe.

361. Lolium multiflorum Lam. Awned Darnel. Hamilton, Madison. From Europe.

362. Lolium temulentum L. Poison Darnel. No specimens. From Europe.

363. Agropyron repens (L.) Beauv. Couch-grass. Rather general; no specimens from the southeastern part of the state. From Europe.

364. Agropyron caninum (L.) R. & S. Awned Wheat-grass. Portage County. From Europe.

365. Triticum vulgare L. Wheat. Erie, Belmont, Harrison, Tuscarawas, Morrow, Fayette, Madison, Preble, Franklin. Escaped from cultivation.

366. Secale cereale L. Rye. Erie, Morrow, Franklin, Scioto. Escaped.

367. Elymus virginicus L. Virginia Wild-rye. General.

368. Elymus hirsutiglumis Scrib. & Sm. Strict Wild-rye. Ottawa, Huron.

369. Elymus canadensis L. Nodding Wild-rye. Rather general.

370. Elymus striatus Willd. Slender Wild-rye. Wayne, Erie, Auglaize.

371. Hystrix hystrix (L.) Millsp. Bottle-brush-grass. General.

372. Hordeum vulgare L. Common Barley. Franklin, Tuscarawas, Portage. Escaped from cultivation.

373. Hordeum distichum L. Two-rowed Barley. Escaped in Lake County.

374. Hordeum nodosum L. Meadow Barley. Hamilton County.

375. Hordeum jubatum L. Squirrel-tail Barley. From Lake to Lucas County; also in Franklin, Madison, Greene, Allen, Defiance, and Williams. Naturalized from the West.

376. Spartina michauxiana Hitch. Tall Slough-grass. Rather general; no specimens from the central eastern nor from the southwestern counties.

377. Beckmannia erucaeformis (L.) Host. Beckmannia. Cuyahoga County.

378. Capriola dactylon (L.) Ktz. Bermuda-grass. No specimens. From Europe.

379. Eleusine indica (L.) Gaert. Yard-grass. General. Naturalized.

380. Atheropogon curtipendulus (Mx.) Fourn. Tall Gramma-grass. Lake, Erie, Ottawa, Franklin, Adams.

381. Bouteloua hirsuta Lag. Hairy Mesquite-grass. Waifs in Franklin County.

382. Bouteloua oligostachya (Nutt.) Torr. Smooth Mesquite-grass. Waifs in Franklin County.

383. Sporobolus asper (Mx.) Kunth. Longleaf Rush-grass. Lake, Erie, Franklin.

384. Sporobolus vaginaeflorus (Torr.) Wood. Sheathed Rush-grass. Auglaize, Madison, Warren, Vinton, Athens.

385. Sporobolus neglectus Nash. Small Rush-grass. Cuyahoga, Wayne, Huron, Auglaize.

386. Sporobolus cryptandrus (Torr.) Gr. Sand Dropseed. Lucas, Ottawa, Erie, Lorain.

387. Sporobolus heterolepis Gr. Northern Dropseed. Franklin, Madison, Champaign.

388. Calamagrostis canadensis (Mx.) Beauv. Bluejoint Reed-grass. Northern Ohio, as far south as Stark, Franklin, and Auglaize Counties.

389. Calamagrostis cinnoides (Muhl.) Scrib. Nuttall's Reed-grass. No specimens.

390. Agrostis alba L. Red-top. General. From Europe.

391. Agrostis schweinitzii Trin. Thin Bent-grass. Rather general.

392. Agrostis hyemalis (Walt.) B. S. P. Rough Silk-grass. Rather general, but no specimens from the northwestern nor southeastern counties.

393. Apera spica-venti (L.) Beauv. Silky Windlestraw. From Europe. Lake County.

394. Cinna arundinacea L. Wood Reed-grass. General.

395. Ammophila arenaria (L.) Link. Sand Beech-grass. Erie County.

396. Alopecurus geniculatus L. Marsh Foxtail. Lake, Ottawa, Crawford, Auglaize, Madison, Franklin, Perry. Introduced.

397. Alopecurus pratensis L. Meadow Foxtail. No specimens. From Europe.

398. Heleochloa schoenoides (L.) Host. Cat-tail-grass. Greene County. From Europe.

399. Phleum pratense L. Timothy. General. From Europe.

400. Muhlenbergia sobolifera (Muhl.) Trin. Rock Muhlenbergia. Highland, Wayne.

401. Muhlenbergia mexicana (L.) Trin. Mexican Muhlenbergia. General.

402. Muhlenbergia racemosa (Mx.) B. S. P. Marsh Muhlenbergia. Summit, Wayne, Huron, Wyandot, Champaign, Licking.

403. Muhlenbergia umbrosa Scribn. Wood Muhlenbergia. Cuyahoga, Champaign.

404. Muhlenbergia tenuiflora (Willd.) B. S. P. Slender Muhlenbergia. Portage, Wayne, Fairfield, Madison, Greene.

405. Muhlenbergia schreberi Gmel. Spreading Muhlenbergia. General.

406. Brachyelytrum erectum (Schreb.) Beauv. Brachyelytrum. Cuyahoga, Portage, Lorain, Wayne, Highland, Franklin, Madison, Hocking, Adams.

407. Milium effusum L. Tall Millet-grass. Lake, Cuyahoga, Lorain, Wayne, Stark.

408. Oryzopsis racemosa (Sm.) Ricker. Black-fruited Mountain-rice. Geauga, Summit, Erie, Greene, Highland.

409. Stipa spartea Trin. Porcupine-grass. Erie County, where it occurs on Cedar Point.

410. Aristida dichotoma Mx. Poverty-grass. Scioto, Vinton, Fairfield.

411. Aristida oligantha Mx. Few-flowered Triple-awned-grass. Cuyahoga County.

412. Aristida gracilis Ell. Slender Triple-awned-grass. Hamilton, Clermont, Athens, Erie, Cuyahoga.

413. Aristida purpurascens Poir. Purplish Triple-awned-grass. Wood, Fulton.

414. Savastana odorata (L.) Scrib. Vanilla-grass. Trumbull, Madison, Pickaway.

415. Phalaris arundinacea L. Reed Canary-grass. Rather general; no specimens from the northwestern nor from the southeastern counties.

416. Phalaris canariensis L. Canary-grass. Montgomery, Hamilton. From Europe.

417. Anthoxanthum odoratum L. Sweet Vernal-grass. Ashtabula, Cuyahoga, Summit, Mahoning, Wayne, Franklin. From Europe.

418. Anthoxanthum puelii Lec. & Lamotte. Long-awned Vernal-grass. Hamilton County. A native of Europe.

Subfamily, PANICATAE.

419. Panicum agrostoides Spreng. Agrostis-like Panic-grass. Erie County.

420. Panicum stipitatum Nash. Long Panic-grass. Northeastern Ohio to Lorain, Fairfield, and Columbiana Counties.

421. Panicum virgatum L. Tall, smooth Panic-grass. General.

422. Panicum dichotomiflorum Mx. Spreading Panic-grass. General.

423. Panicum miliaceum L. Millet Panic-grass. Lawrence, Richland, Erie. Introduced.

424. Panicum capillare L. Tumble Panic-grass. General and abundant.

425. Panicum gattingeri Nash. Gattinger's Panic-grass. Rather general.

426. Panicum flexile (Gatt.) Scrib. Wiry Panic-grass. Adams, Champaign, Madison, Franklin, Erie, Cuyahoga.

427. Panicum philadelphicum Bernh. Philadelphia Panic-grass. Ottawa County.

428. Panicum depauperatum Muhl. Starved Panic-grass. Cuyahoga County.

429. Panicum linearifolium Scrib. Linear-leaf Panic-grass. Rather general.

430. Panicum werneri Scrib. Werner's Panic-grass. Lake, Cuyahoga, Franklin, Athens.

431. Panicum bicknellii Nash. Bicknell's Panic-grass. Gallia County.

432. Panicum sphaerocarpon Ell. Round-fruited Panic-grass. Cuyahoga, Summit, Trumbull, Hocking, Scioto.

433. Panicum polyanthes Schultes. Many-flowered Panic-grass. Fairfield, Hocking, Jackson.

434. Panicum dichotomum L. Forked Panic-grass. Rather general; no specimens fro mthe northwestern counties.

435. Panicum microcarpon Muhl. Small-fruited Panic-grass. Cuyahoga, Lorain, Eriè, Fairfield, Hocking, Jackson, Adams.

436. Panicum boreale Nash. Northern Panic-grass. Fulton County.

437. Panicum lindheimeri Nash. Lindheimer's Panic-grass. Ashtabula, Hocking.

438. Panicum huachucae Ashe. Hairy Panic-grass. General.

439. Panicum villosissimum Nash. Villous Panic-grass. Cuyahoga, Erie, Licking.

440. Panicum implicatum Scrib. Slender-stemmed Panic-grass. Gallia County.

441. Panicum tsugetorum Nash. Hemlock Panic-grass. Defiance, Summit.

442. Panicum leibergii (Vasey) Scrib. Leiberg's Panic-grass. No specimens.

443. Panicum scribnerianum Nash. Scribner's Panic-grass. Cuyahoga, Erie, Wood, Franklin.

444. Panicum xanthophysum Gr. Slender Panic-grass. Lake County.

445. Panicum ashei Pear. Ashe's Panic-grass. Cuyahoga, Lake, Trumbull, Fairfield.

446. Panicum commutatum Schultes. Variable Panic-grass. Lawrence, Gallia, Fairfield, Wayne.

447. Panicum latifolium L. Broad-leaf Panic-grass. General.

448. Panicum boscii Poir. Bosc's Panic-grass. Warren, Adams, Jackson, Belmont.

448a. Panicum boscii molle (Vas.) Hitch. & Ch. Hamilton, Lawrence, Cuyahoga.

449. Panicum clandestinum L. Hispid Panic-grass. General.

450. Leptaloma cognatum (Schultes) Chase. Fall Witch-grass. From the West. Lake County.

451. Syntherisma filiforme (L.) Nash. Slender Crab-grass. No specimens.

452. Syntherisma ischaemum (Schreb.) Nash. Small Crab-grass. Lorain, Wayne, Auglaize, Fairfield. From Europe.

453. Syntherisma sanguinale (L.) Dulac. Large Crab-grass. General. Naturalized.

454. Echinochloa crus-galli (L.) Beauv. Common Barnyard-grass. General and abundant. Naturalized from Europe.

455. Echinochloa walteri (Pursh.) Nash. Marsh Cockspur-grass. Erie, Lorain, Shelby, Auglaize, Licking.

456. Paspalum muhlenbergii Nash. Muhlenberg's Paspalum. Cuyahoga, Erie, Warren, Hamilton, Scioto, Guernsey.

457. Chaetochloa verticillata (L.) Scrib. Verticillate Foxtail-grass. From Europe. Cuyahoga, Wayne, Jefferson, Franklin, Ross, Montgomery, Warren, Hamilton.

458. Chaetochloa glauca (L.) Scrib. Yellow Foxtail-grass. General. From Europe.

459. Chaetochloa viridis (L.) Scrib. Green Foxtail-grass. General. Naturalized from Europe.

460. Chaetochloa italica (L.) Scrib. Italian Millet. Rather general. Escaped.

461. Cenchrus tribuloides L. Sandbur-grass. Lucas, Wood, Ottawa, Erie, Lorain, Cuyahoga, Franklin, Highland, Gallia.

462. Homalocenchrus virginicus (Willd.) Britt. Virginia Cut-grass. Rather general.

463. Homalocenchrus oryzoides (L.) Poll. Rice Cut-grass. Rather general.

464. Zizania aquatica L. Wild Rice. Erie, Licking, Perry, Hocking.

465. Holcus halapense L. Johnson-grass. Native of Europe. Cuyahoga, Erie, Franklin, Madison

466. Holcus sorghum L. Common Sorghum. Volunteer in Adams County.

467 Sorghastrum nutans (L.) Nash. Indian-grass. Ashtabula, Cuyahoga, Erie, Wyandot, Auglaize, Franklin, Madison, Adams.

468. Miscanthus sinensis Anderss. Chinese Plume-grass. An escape in Lake County.

469. Andropogon furcatus Muhl. Big Bluestem. Rather general.

470. Andropogon virginicus L. Virginia Beard-grass. Gallia, Jackson, Meigs, Athens, Vinton, Hocking, Fairfield, Belmont.

471. Andropogon scoparius Mx. Little Bluestem. (Schizachyrium scoparium (Mx.) Nash). Rather general.

472. Coix lacryma L. Job's-tears. Persistent in Franklin County.

473. Zea mays L. Indian Corn. Spontaneous in Brown, Adams, Scioto, Fayette, Monroe, and Hancock Counties.

Subclass, LILIIFLORAE.

Order, *Liliales.*

Liliaceae. Lily Family.

Subfamily, DRACAENATAE.

474. Yucca filamentosa L. Adam's-needle. Escaped in Franklin County

Subfamily, LILIATAE.

475. Lilium superbum L. Turk's-cap Lily. Erie County—Moseley herbarium).
476. Lilium canadense L. Canada Lily. General.
477. Lilium philadelphicum L. Philadelphia Lily. Fulton, Lucas, Sandusky, (Erie County—Moseley herbarium).
478. Lilium umbellatum Pursh. Western Red Lily. Stark County.
479. Erythronium americanum Ker. Yellow Dog-tooth Lily. General.
480. Erythronium albidum Nutt. White Dog-tooth Lily. General.
481. Hemerocallis fulva L. Common Day-lily. General. Escaped from cultivation.
482. Allium tricoccum Ait. Wild Leek. West central part of state to Delaware and Franklin; also in Lorain, Cuyahoga, and Medina Counties.
483. Allium vineale L. Field Garlic. Franklin, Harrison. From Europe.
484. Allium cepa L. Common Onion. Cultivated. Sometimes persistent.
485. Allium canadense L. Meadow Garlic. General.
486. Allium cernuum Roth. Nodding Onion. General.
487. Quamasia hyacinthina (Raf.) Britt. Wild Hyacinth. General, but rare in eastern Ohio.
488. Ornithogalum umbellatum L. Star-of-Bethlehem. From Europe. Montgomery, Miami, Gallia, Franklin, Auglaize.
489. Muscari botryoides (L.) Mill. Grape-hyacinth. From Europe. Montgomery, Lake.
490. Aletris farinosa L. Colic-root. In counties bordering Lake Erie.

Subfamily, MELANTHATAE.

491. Uvularia sessilifolia L. Sessile-leaf Bellwort. Lucas, Cuyahoga, Portage, Summit, Mahoning, Gallia.

492. Uvularia grandiflora Sm. Large-flowered Bellwort. General.
493. Uvularia perfoliata L. Perfoliate Bellwort. General.
494. Melanthium virginicum L. Virginia Bunchflower. Richland, Wayne.
495. Veratrum woodii Robb. Wood's False-hellebore. Auglaize County.
496. Veratrum viride Ait. American False-hellebore. Ashtabula County.
497. Anticlea elegans (Pursh) Rydb. Glaucous Anticlea. Champaign, Portage, Stark, Highland, Ottawa.
498. Stenanthium robustum Wats. Stout Stenanthium. Fairfield County.
499. Chamaelirium luteum (L.) Gr. Chamaelirium. Northeastern Ohio to Licking County; also in Lawrence County.
500. Triantha glutinosa (Mx.) Baker. Glutinous Triantha. Stark, Champaign.

Subfamily, TRILLIATAE.

501. Trillium grandiflorum (Mx.) Salisb. Large-flowered Trillium. General.
502. Trillium erectum L. Ill-scented Trillium. General.
503. Trillium cernuum L. Nodding Trillium. Auglaize, Champaign, Medina.
504. Trillium declinatum (Gr.) Gleason. Drooping Trillium. No specimens.
505. Trillium undulatum Willd. Painted Trillium. Ashtabula County.
506. Trillium nivale Ridd. Early Trillium. Miami, Clark, Greene, Franklin.
507. Trillium sessile L. Sessile Trillium. General.
508. Trillium recurvatum Beck. Prairie Trillium. Auglaize, Hamilton.
509. Medeola virginiana L. Indian Cucumber-root. General.

Subfamily, CONVALLARIATAE.

510. Streptopus amplexifolius (L.) DC. Clasping-leaf Twisted-stalk. Reported for Ohio.
511. Disporum lanuginosum (Mx.) Nich. Hairy Disporum. Eastern half of state; also in Adams County.

512. Polygonatum commutatum (R. & S.) Dietr. Smooth Solomon's-seal. General.
513. Polygonatum biflorum (Walt.) Ell. Hairy Solomon's-seal. General.
514. Vagnera racemosa (L.) Mor. Panicled False Solomon's-seal. General.
515. Vagnera stellata (L.) Mor. Stellate False Solomon's-seal. Rather general.
516. Vagnera trifolia (L.) Mor. Three-leaf False Solomon's-seal. Fulton, Lorain.
517. Unifolium canadense (Desf.) Greene. False Lily-of-the-valley. General.
518. Clintonia borealis (Ait.) Raf. Yellow Clintonia. Ashtabula County.
519. Clintonia umbellulata (Mx.) Torr. White Clintonia. Harrison, Portage, Wayne.
520. Convallaria majalis L. Lily-of-the-valley. Escaped in Franklin County.
521. Asparagus officinalis L. Asparagus. General. Introduced from Europe.

Smilaceae. Smilax Family.

522. Smilax ecirrhata (Engel.) Wats. Upright Smilax. Erie, Ottawa, Wood, Hardin, Preble, Clinton, Brown, Fairfield.
523. Smilax herbacea L. Common Carrion-flower. General.
524. Smilax pseudo-china L. Long-stalked Greenbrier. Erie, Vinton, Brown.
525. Smilax hispida Muhl. Hispid Greenbrier. General.
526. Smilax glauca Walt. Glaucous Greenbrier. Southeastern part of the state to Clermont, Warren, Fairfield, Knox, and Summit Counties; also in Lucas County.
527. Smilax rotundifolia L. Roundleaf Greenbrier. Lawrence, Hocking, Fairfield, Licking, Belmont, Lorain, Cuyahoga.

Pontederiaceae. Pickerel-weed Family.

528. Pontederia cordata L. Pickerel-weed. From Licking and Perry northeast; also in Defiance, Fulton, Lucas, and Erie.
529. Heteranthera dubia (Jacq.) MacM. Water-stargrass. Rather general.

Commelinaceae. Spiderwort Family.

530. Tradescantia reflexa Raf. Reflexed Spiderwort. Ashtabula, Erie, Mahoning, Richland, Coshocton, Licking, Franklin, Auglaize.
531. Tradescantia virginiana L. Virginia Spiderwort. General as far north as Auglaize and Stark.
532. Tradescantia pilosa Lehm. Zigzag Spiderwort. Hamilton, Clermont, Montgomery.
533. Commelina virginica L. Virginia Day-flower. Clermont, Montgomery, Lake.

Juncaceae. Rush Family.

534. Juncus effusus L. Common Rush. General.
535. Juncus balticus Willd. Baltic Rush. Erie County.
536. Juncus gerardi Lois. Gerard's Rush. Cuyahoga County.
537. Juncus dudleyi Wieg. Dudley's Rush. Montgomery, Clinton, Champaign, Delaware, Licking, Tuscarawas.
538. Juncus tenuis Willd. Slender Rush. General.
539. Juncus bufonius L. Toad Rush. Williams, Lucas, Lorain, Licking.
540. Juncus monostichus Bartl. One-ranked Rush. Trumbull County.
541. Juncus aristulatus Mx. Small-headed Grass-leaf Rush. Fairfield, Wayne, Summit.
542. Juncus marginatus Rostk. Grass-leaf Rush. Cuyahoga County.
543. Juncus alpinus Vill. Richardson's Rush. Cuyahoga County.
544. Juncus articulatus L. Jointed Rush. Cuyahoga County.
545. Juncus torreyi Cov. Torrey's Rush. Adams, Madison, Wood, Erie, Cuyahoga.
546. Juncus nodosus L. Knotted Rush. Madison, Franklin, Cuyahoga, Erie.
547. Juncus brachycephalus (Engelm.) Buch. Small-headed Rush. Erie, Cuyahoga, Franklin, Madison, Champaign.
548. Juncus acuminatus Mx. Sharp-fruited Rush. General.
549. Juncus canadensis J. Gay. Canada Rush. Cuyahoga, Geauga, Licking, Madison, Auglaize.
550. Juncus scirpoides Lam. Scirpus-like Rush. Erie County.
551. Juncoides carolinae (Wats.) Ktz. Hairy Wood-rush. Lucas, Cuyahoga, Trumbull, Mahoning, Hocking.
552. Juncoides campestre (L.) Ktz. Common Wood-rush. General.

Xyridaceae. Yellow-eyed-grass Family.

553. Xyris flexuosa Muhl. Slender Yellow-eyed-grass. Portage, Geauga.

Eriocaulaceae. Pipewort Family.

554. Eriocaulon septangulare With. Seven-angled Pipewort. Summit County.

Order, *Iridales.*

Amaryllidaceae. Amaryllis Family.

555. Manfreda virginica (L.) Salisb. False Aloe. Lawrence County.
556. Hypoxis hirsuta (L.) Cov. Yellow Stargrass. General.

Iridaceae. Iris Family.

557. Iris versicolor L. Large Blue-flag. General.
558. Iris cristata Ait. Crested Dwarf Iris. Lawrence, Adams, Scioto, Pike, Ross, Jackson, Vinton, Hocking, Cuyahoga, Trumbull.
559. Gemmingia chinensis (L.) Ktz. Blackberry-lily. From Asia. Franklin County.
560. Crocus verna L. Crocus. Escaped in Lake County.
561. Sisyrinchium angustifolium Mill. Pointed Blue-eyed-grass. Rather general.
562. Sisyrinchium graminoides Bickn. Stout Blue-eyed-grass. General.

Dioscoreaceae. Yam Family.

563. Dioscorea villosa L. Wild Yam. General.

563.1. Dioscorea bulbifera L. Air Potato (Yam). Escaped from gardens in Madison County.

Order, *Orchidales.*

Orchidaceae. Orchid Family.

Subfamily, CYPRIPEDIATAE.

564. Cypripedium reginae Walt. Showy Lady's-slipper. Fulton, Champaign, Lucas, Geauga, Portage, Muskingum.
565. Cypripedium candidum Willd. White Lady's-slipper. Wyandot, Erie, Montgomery.

566. Cypripedium parviflorum Salisb. Yellow Lady's-slipper. General.

567. Fissipes acaulis (Ait.) Small. Moccasin-flower. Medina, Portage, Hocking, Fairfield, Stark, Cuyahoga.

Subfamily, ORCHIDATAE.

568. Galeorchis spectabilis (L.) Rydb. Showy Orchis. General.

569. Perularia flava (L.) Farw. Tubercled Orchis. General.

570. Coeloglossum bracteatum (Willd.) Parl. Long-bracted Orchis. Lucas, Lorain, Medina, Portage, Franklin, Butler, Auglaize.

571. Gymnadeniopsis clavellata (Mx.) Rydb. Green Wood-orchis. Geauga, Trumbull, Portage, Summit, Richland, Licking, Champaign.

572. Limnorchis hyperborea (L.) Rydb. Tall Bog-orchis. Stark County.

573. Lysias orbiculata (Pursh.) Rydb. Large Roundleaf Orchis. Cuyahoga, Geauga, Wayne.

574. Lysias hookeriana (Gr.) Rydb. Hooker's Orchis. Medina County.

575. Blephariglottis ciliaris (L.) Rydb. Yellow Fringed-orchis. Fulton County.

576. Blephariglottis blephariglottis (Willd.) Rydb. White Fringed-orchis. Geauga, Portage.

577. Blephariglottis lacera (Mx.) Farw. Ragged Fringed-orchis. Cuyahoga, Portage, Crawford, Richland, Wayne, Holmes, Stark, Licking, Fairfield.

578. Blephariglottis leucophaea (Nutt.) Farw. Prairie Fringed-orchis. Auglaize County.

579. Blephariglottis psycodes (L.) Rydb. Smaller Purple Fringed-orchis. Rather general.

579a. Blephariglottis psycodes grandiflora (Bigel.). Portage County (R. J. Webb).

580. Blephariglottis peramoena (Gr.) Rydb. Fringeless Purple Orchis. Perry, Gallia, Clermont, Wayne, Hocking.

581. Pogonia ophioglossoides (L.) Ker. Rose Pogonia. Lucas, Cuyahoga, Geauga, Ashland, Portage, Richland, Licking, Lorain, Holmes.

582. Isotria verticillata (Willd.) Raf. Whorled Isotria. Defiance, Cuyahoga, Geauga, Medina, Coshocton, Fairfield.

583. Triphora trianthophora (Sw.) Rydb. Nodding Triphora. Huron, Cuyahoga, Wayne, Summit, Stark, Licking, Franklin, Ross, Clark.

584. Arethusa bulbosa L. Arethusa. Licking, Portage.

585. Limodorum tuberosum L. Limodorum. Fulton, Lucas, Erie, Geauga, Portage, Summit, Ashland, Stark, Clark, Fairfield, Licking, Wayne.

586. Ibidium strictum (Rydb.) House. Hooded Lady's-tresses. Ashtabula County.

587. Ibidium plantagineum (Raf.) House. Broadleaf Lady's-tresses. Medina, Portage.

588. Ibidium cernuum (L.) House. Nodding Lady's-tresses. Erie, Lorain, Cuyahoga, Medina, Portage, Stark, Lake, Licking.

589. Ibidium ovale (Lindl.) House. Small-flowered Lady's-tresses. No specimens.

590. Ibidium praecox (Walt.) House. Grass-leaf Lady's-tresses. Wayne County.

591. Ibidium beckii (Lindl.) House. Little Lady's-tresses. Fairfield County.

592. Ibidium gracile (Bigel.) House. Slender Lady's-tresses. Erie, Cuyahoga, Lake, Licking, Muskingum, Fairfield, Hocking, Adams, Gallia, Morgan.

593. Peramium pubescens (Willd.) MacM. Downy Rattlesnake-plantain. Adams, Hocking, Lake, Fairfield, Highland, Wayne, Noble, Portage.

594. Malaxis unifolia Mx. Green Addermouth. Fairfield, Hocking, Wayne.

595. Liparis liliifolia (L.) Rich. Large Twayblade. Portage, Franklin, Fairfield, Clark, Wayne.

596. Liparis loeselii (L.) Rich. Fen Twayblade. Champaign, Cuyahoga, Summit, Stark.

597. Tipularia unifolia (Muhl.) B. S. P. Crane-fly Orchis. Medina, (Lorain, Cuyahoga—Oberlin College Herbarium).

598. Aplectrum hyemale (Muhl.) Torr. Putty-root. General.

599. Corallorrhiza corallorrhiza (L.) Karst. Early Coral-root. No specimens.

600. Corallorrhiza maculata Raf. Large Coral-root. Erie, Huron, Cuyahoga, Wayne, Fairfield, Franklin, Gallia, Noble.

601. Corallorrhiza wisteriana Conrad. Wister's Coral-root. Hamilton, Lawrence.
602. Corallorrhiza odontorhiza (Willd.) Nutt. Small-flowered Coral-root. Erie, Cuyahoga, Stark, Licking, Fairfield, Madison.

Class, DICOTYLAE. Dicotyls.

Subclass, THALAMIFLORAE.

Order, *Ranales*.

Magnoliaceae. Magnolia Family.

603. Magnolia acuminata L. Cucumber Magnolia. Eastern half of state, west to Lorain and Madison Counties.
604. Magnolia tripetala L. Umbrella Magnolia. Scioto County.
605. Liriodendron tulipifera L. Tuliptree. General.

Anonaceae. Custard-apple Family.

606. Asimina triloba (L.) Dunal. Papaw. General.

Ranunculaceae. Crowfoot Family.

607. Ranunculus abortivus L. Kindney-leaf Crowfoot. General and abundant.
608. Ranunculus micranthus Nutt. Rock Crowfoot. Clermont County.
609. Ranunculus sceleratus L. Celery-leaf Crowfoot. Rather general.
610. Ranunculus recurvatus Poir. Hooked Crowfoot. General and abundant.
611. Ranunculus acris L. Tall Buttercup. Rather general except in the southern part. From Europe.
612. Ranunculus bulbosus L. Bulbous Buttercup. Columbiana County. From Europe.
613. Ranunculus pennsylvanicus L. f. Bristly Buttercup. Lucas, Ottawa, Cuyahoga, Lake, Wayne, Licking, Fairfield, Perry.
614. Ranunculus repens L. Creeping Buttercup. Scioto, Columbiana. From Europe.
615. Ranunculus septentrionalis Poir. Swamp Buttercup. General and abundant.
616. Ranunculus hispidus Mx. Hispid Buttercup. General.
617. Ranunculus fascicularis Muhl. Tufted Buttercup. Lucas, Ottawa, Cuyahoga.

618. Ranunculus arvensis L. Corn Crowfoot. No specimens. From Europe.
619. Ranunculus obtusiusculus Raf. Lance-leaf Buttercup. Jackson, Franklin, Licking, Erie, Lorain, Cuyahoga, Lake, Ashtabula.
620. Ranunculus delphinifolius Torr. Yellow Water-crowfoot. Northwestern fourth of state to Huron and Madison Counties; also in Ashtabula County.
621. Ficaria ficaria (L.) Karst. Golden-cup. Lake County From Europe.
622. Batrachium trichophyllum (Chaix.) Schultz. White Water-crowfoot. Rather general.
623. Batrachium circinatum (Sibth.) Rchb. Circinate White Water-crowfoot. Licking, Defiance.
624. Trollius laxus Salisb. American Globe-flower. Columbiana, Stark.
625. Helleborus viridis L. Green Hellebore. Gallia, Miami, Franklin, Stark. From Europe.
626. Nigella damascena L. Love-in-a-mist. Hamilton County. (Erie County—Moseley Herbarium.) Escaped from gardens.
627. Coptis trifolia (L.) Salisb. Gold-thread. Defiance, Portage, Summit, Stark, Geauga.
628. Aquilegia canadensis L. Wild Columbine. General.
629. Aquilegia vulgaris L. European Columbine. Escaped in Fulton County.
630. Aconitum noveboracense Gr. New York Monkshood. Summit, Portage.
631. Delphinium tricorne Mx. Dwarf Larkspur. Southern half of the state; also in Columbiana County
632. Delphinium exaltatum Ait. Tall Larkspur. Stark, Franklin.
633. Delphinium ajacis L. Garden Larkspur. General. Naturalized from Europe.
634. Anemone cylindrica Gr. Long-fruited Anemone. Wood, Ottawa, Erie.
635. Anemone virginiana L. Virginia Anemone. General.
636. Anemone canadensis L. Canada Anemone. General.
637. Anemone quinquefolia L. Wind-flower. General except southern and southeastern parts of the state.
638. Hepatica hepatica (L.) Karst. Roundlobed Liver-leaf. General.

639. Hepatica acutiloba DC. Sharplobed Liver-leaf. General.
640. Clematis virginiana L. Virginia Virgin's-bower. General.
641. Viorna viorna (L.) Small. Leather-flower. Southern half of Ohio; also in Auglaize County.
642. Caltha palustris L. Marsh-marigold. General.
643. Hydrastis canadensis L. Golden-seal. General.
644. Actaea rubra (Ait.) Willd. Red Baneberry. Erie, Sandusky.
645. Actaea alba (L.) Mill. White Baneberry. General.
646. Cimicifuga racemosa (L.) Nutt. Black Cohosh. Eastern half of state to Erie, Fairfield, and Clermont Counties.
647. Syndesmon thalictroides (L.) Hoffm. Rue-anemone. General and abundant.
648. Isopyrum biternatum (Raf.) T. & G. False Rue-anemone. Southwestern fourth of state; also in Cuyahoga County.
649. Thalictrum dioicum L. Early Meadow-rue. General.
650. Thalictrum dasycarpum Fisch & Lall. Purplish Meadow-rue. General.
651. Thalictrum polygamum Muhl. Fall Meadow-rue. Rather general.

Parnissiaceae. Grass-of-Parnassus Family.

652. Parnassia caroliniana Mx. Carolina Grass-of-Parnassus. Rather general.

Ceratophyllaceae. Hornleaf Family.

653. Ceratophyllum demersum L. Hornleaf. General.

Berberidaceae. Barberry Family.

654. Podophyllum peltatum L. May-apple. General and abundant.
655. Jeffersonia diphylla (L.) Pers. Twinleaf. General.
656. Caulophyllum thalictroides (L.) Mx. Blue Cohosh. General.
657. Berberis vulgaris L. Common Barberry. Escaped rather generally.
658. Odostemon aquifolium (Pursh.) Rydb. Tailing Mahonia. Escaped in Lake County.

Menispermaceae. Moonseed Family.

659. Menispermum canadense L. Moonseed. General and abundant.

Lauraceae. Laurel Family.

660. Sassafras sassafras (L.) Karst. Sassafras. General.
661. Benzoin aestivale (L.) Nees. Spicebush. General.

Order, *Sarraceniales.*

Sarraceniaceae. Pitcher-plant Family.

662. Sarracenia purpurea L. Pitcher-plant. Geauga, Summit, Ashtabula, Wayne, Richland, Defiance, Williams.

Droseraceae. Sundew Family.

663. Drosera rotundifolia L. Roundleaf Sundew. Licking, Wayne, Portage, Stark, Geauga, Ashtabula.
664. Drosera intermedia Hayne. Spatulate Sundew. Wayne County.

Order, *Brassicales.*

Papaveraceae. Poppy Family.

665. Papaver somniferum L. Opium Poppy. No specimens. From Europe.
666. Papaver rhoeas L. Field Poppy. Ashtabula County. From Europe.
667. Papaver dubium L. Corn Poppy. No specimens. From Europe.
668. Papaver argemone L. Rough-fruited Poppy. (Erie County-Moseley Herbarium.) From Europe.
669. Argemone mexicana L. Mexican Prickly-poppy. Franklin County. From tropical America.
670. Sanguinaria canadensis L. Bloodroot. General and abundant.
671. Stylophorum diphyllum (Mx.) Nutt. Yellow Poppy. Southern half of state.
672. Macleya cordata (Willd.) R. Br. Macleya. Escaped in Madison and Franklin Counties.
673. Chelidonium majus L. Celandine. General. From Europe.

Fumariaceae. Fumitory Family.

674. Bicuculla cucullaria (L.) Millsp. Dutchman's-breeches. General.
675. Bicuculla canadensis (Goldie) Millsp. Squirrel-corn. General.

676. Adlumia fungosa (Ait.) Greene. Climbing Fumitory. Lorain, Cuyahoga, Lake, Summit, Belmont.

677. Capnoides sempervirens (L.) Borck. Pink Corydalis. Fairfield, Knox, Portage.

678. Capnoides flavulum (Raf.) Ktz. Pale Corydalis. Southwestern fourth of state; also in Ottawa and Erie Counties.

679. Capnoides aureum (Willd.) Ktz. Golden Corydalis. (Ottawa County—Moseley Herbarium.)

680. Fumaria officinalis L. Common Fumitory. Columbiana County. From Europe.

681. Fumaria parviflora Lam. Small-flowered Fumitory. Introduced in Lake County.

Brassicaceae. Mustard Family.

682. Berteroa incana (L.) DC. Hoary Berteroa. No specimens. From Europe.

683. Koniga maritima (L.) R. Br. Sweet Alyssum. Erie County. Escaped.

684. Alyssum alyssoides L. Yellow Alyssum. Sandusky County. From Europe.

685. Draba verna L. Vernal Whitlow-grass. Southern half of state; also in Portage County. From Europe.

686. Draba caroliniana Walt. Carolina Whitlow-grass. Adams, Clark, Erie, Ottawa.

687. Camelina sativa (L.) Crantz. Common False-flax. Sandusky, Auglaize, Miami, Montgomery, Franklin. From Europe.

688. Camelina microcarpa Andrz. Small-fruited False-flax. Clark County. From Europe.

689. Bursa bursa-pastoris (L.) Britt. Shepherd's-purse. General and abundant. From Europe.

690. Neslia paniculata (L.) Desv. Ball-mustard. Escaped in Lake County.

691. Armoracia armoracia (L.) Britt. Horseradish. General. From Europe.

692. Neobeckia aquatica (Eat.) Britt. Lake Water-cress. Lucas, Coshocton, Licking, Perry, Madison.

693. Sisymbrium nasturtium-aquaticum L. True Water-cress. Rather general. From Europe.

694. Radicula hispida (Desv.) Britt. Hispid Yellow-cress. Monroe, Shelby, Logan, Ottawa, Erie, Huron, Cuyahoga, Summit.

695. Radicula palustris (L.) Moench. Marsh Yellow-cress. General.

696. Radicula sylvestris (L.) Druce. Creeping Yellow-cress. Lucas, Erie, Cuyahoga. From Europe.

697. Lepidium ruderale L. Roadside Peppergrass. No specimens. From Europe.

698. Lepidium virginicum L. Virginia Peppergrass. General and abundant.

699. Lepidium densiflorum Schrad. Wild Peppergrass. Auglaize, Champaign, Franklin, Fayette, Wayne, Lorain, Cuyahoga, Lake.

700. Lepidium draba L. Hoary Peppergrass. Lucas County. From Europe.

701. Lepidium campestre (L.) R. Br. Field Peppergrass. Rather general. From Europe.

702. Carara didyma (L.) Britt. Lesser Wart-cress. Lake County. Escaped.

703. Thlaspi arvense L. Field Penny-cress. Cuyahoga, Henry. From Europe.

704. Myagrum perfoliatum L. Myagrum. Lake County. From Europe.

705. Alliaria alliaria (L.) Britt. Garlic Mustard. (Erie County—Moseley Herbarium.) From Europe.

706. Sophia pinnata (Walt.) Howell. Pinnate Tanzy-mustard. Hamilton, Montgomery, Miami, Ottawa, Jackson.

707. Sophia incisa (Eng.) Greene. Western Tanzy-mustard. Miami, Portage.

708. Cheirinia cheiranthoides (L.) Link. Worm-seed Mustard. Hamilton, Lucas, Lake, Portage.

709. Cheirinia repanda (L.) Link. Repand Cheirinia. Logan, Erie. From Europe.

710. Cheirinia aspera (DC.) Britt. Western Cheirinia. Franklin County.

711. Erysimum officinale L. Hedge-mustard. General and abundant. From Europe.

712. Norta altissima (L.) Britt. Tall Hedge-mustard. Lake, Cuyahoga, Erie, Ottawa, Wayne, Jackson, Greene, Belmont, Portage. From Europe.

713. Norta irio (L.) Britt. Longleaf Hedge-mustard. Portage County. Introduced.

714. Conringia orientalis (L.) Dum. Hare's-ear Mustard. Lake, Cuyahoga, Geauga. From Europe.

715. Hesperis matronalis L. Dame's Rocket. Hamilton, Franklin, Portage. From Europe.

716. Arabidopsis thaliana (L.) Britt. Mouse-ear Cress. Clinton, Montgomery, Lucas, Asthtabula. From Europe.

717. Barbarea barbarea (L.) MacM. Yellow Winter-cress. General. From Europe.

718. Barbarea stricta Andrz. Erect Winter-cress. Erie County. From Europe.

719. Barbarea verna (Mill.) Aschers. Early Winter-cress. Belmont, Portage, Harrison, Preble. From Europe.

720. Iodanthus pinnatifidus (Mx.) Steud. Purple Rocket. Rather general.

721. Arabis dentata T. & G. Toothed Rock-cress. Rather general.

722. Arabis patens Sull. Spreading Rock-cress. Franklin County.

723. Arabis hirsuta (L.) Scop. Hairy Rock-cress. Rather general.

724. Arabis glabra (L.) Bernh. Tower Mustard. Hamilton, Franklin, Auglaize, Lucas, Richland, Cuyahoga, Geauga, Belmont.

725. Arabis laevigata (Muhl.) Poir. Smooth Rock-cress. General.

726. Arabis canadensis L. Sickle-pod Rock-grass. General.

727. Arabis virginica (L.) Trel. Virginia Rock-cress. Clermont, Lawrence, Clark.

728. Arabis lyrata L. Lyre-leaf Rock-cress. Pike, Muskingum, Auglaize, Wood, Erie.

729. Arabis drummondii Gr. Drummond's Rock-cress. No specimens.

730. Arabis brachycarpa (T. & G.) Britt. Purple Rock-cress. Erie, Ottawa.

731. Cardamine douglassii (Torr.) Britt. Purple Bitter-cress. General and abundant.

732. Cardamine bulbosa (Schreb.) B. S. P. Bulbous Bitter-cress. General.

733. Cardamine rotundifolia Mx. Roundleaf Bitter-cress. Belmont, Noble.

734. Cardamine pratensis L. Meadow Bitter-cress. Portage County.

735. Cardamine hirsuta L. Hairy Bitter-cress. Lake County.

736. Cardamine pennsylvanica Muhl. Pennsylvania Bitter-cress. General.
737. Cardamine arenicola Britt. Sand Bitter-cress. Lake County.
738. Cardamine parviflora L. Small-flowered Bitter-cress. Lawrence, Hocking, Fairfield, Delaware.
739. Dentaria dipylla Mx. Two-leaf Toothwort. Eastern half of state.
740. Dentaria maxima Nutt. Large Toothwort. No specimens.
741. Dentaria heterophylla Nutt. Slender Toothwort. Auglaize, Hocking, Vinton, Belmont, Clermont.
742. Dentaria laciniata Muhl. Cutleaf Toothwort. General and abundant.
743. Sinapis alba L. White Mustard. Lucas County. From Europe.
744. Sinapis arvensis L. Corn Mustard. General except in the southern part. From Europe.
745. Brassica nigra (L.) Koch. Black Mustard. General and abundant. From Europe.
746. Brassica juncea (L.) Cosson. Indian Mustard. Wayne, Portage. From Asia.
747. Brassica campestris L. Common Turnip. Miami, Auglaize, Franklin, Wayne. From Europe.
748. Brassica napus L. Rape. Franklin County. Introduced.
749. Brassica oleracea L. Cabbage. No specimens. Spontaneous after cultivation.
750. Diplotaxis muralis (L.) DC. Sand Rocket. Cuyahoga County. From Europe.
751. Raphanus raphanistrum L. Wild Radish. Lake County. From Europe.
752. Raphanus sativus L. Garden Radish. Rather general. Spontaneous after cultivation.
753. Cakile edentula (Bigel.) Hook. Sea Rocket. Ashtabula, Lake, Cuyahoga, Erie.

Capparidaceae. Caper Family.

754. Polanisia graveolens Raf. Clammy-weed. Hamilton, Clermont, Warren, Montgomery, Greene, Ross, Monroe, Erie, Cuyahoga.
755. Cleome spinosa L. Spider-flower. Montgomery, Cuyahoga. From tropical America.

Resedaceae. Mignonette Family.

756. Reseda luteola L. Dyer's Mignonette. Licking, Belmont. From Europe.

757. Reseda alba L. White Mignonette. Cuyahoga County. From Europe.

Order, *Geraniales.*

Geraniaceae. Geranium Family.

758. Geranium maculatum L. Wild Crane's-bill. General and abundant.

759. Geranium columbinum L. Long-stalked Crane's-bill. Lake County. From Europe.

760. Geranium carolinianum L. Carolina Crane's-bill. General.

761. Geranium molle L. Dove's-foot Crane's-bill. Madison, Lake. From Europe.

762. Geranium pusillum L. Small-flowered Crane's-bill. Cuyahoga, Lake, Ashtabula. From Europe.

763. Robertiella robertiana (L.) Hanks. Herb-Robert. North central counties.

764. Erodium cicutarium (L.) L'Her. Stork's-bill. Auglaize, Lake. From Europe.

Oxalidaceae. Wood-sorrel Family.

765. Oxalis grandis Small. Great Wood-sorrel. General except in the northwestern part of the state.

766. Oxalis cymosa Small. Tall Wood-sorrel. General and abundant.

767. Oxalis stricta L. Upright Wood-sorrel. General.

768. Oxalis brittoniae Small. Britton's Wood-sorrel. Franklin, Lake.

769. Oxalis rufa Small. Red Wood-sorrel. Franklin, Lake.

770. Oxalis corniculata L. Procumbent Wood-sorrel. Monroe, Franklin, Lake. From tropical America.

771. Oxalis violacea L. Violet Wood-sorrel. Rather general.

Limnanthaceae. False-mermaid Family.

772. Floerkea proserpinacoides Willd. False-mermaid. Rather general.

Linaceae. Flax Family.

773. Linum usitatissimum L. Common Flax. Rather general. Introduced.
774. Linum perenne L. Perennial Flax. Escaped in Lake County.
775. Linum virginianum L. Virginia Flax. Eastern half of state to Erie, Franklin, and Adams Counties.
776. Linum medium (Planch.) Britt. Stiff Flax. Erie County.
777. Linum sulcatum Ridd. Grooved Flax. Erie County.

Balsaminaceae. Jewel-weed Family.

778. Impatiens pallida Nutt. Pale Touch-me-not. General.
779. Impatiens biflora Walt. Spotted Touch-me-not. General and abundant.

Rutaceae. Rue Family.

780. Zanthoxylum americanum Mill. Prickly-ash. General in western Ohio as far east as Huron and Licking Counties.
781. Ptelea trifoliata L. Hoptree. General.

Simarubaceae. Ailanthus Family.

782. Ailanthus gladulosa Desf. Tree-of-heaven. General. Introduced.

Polygalaceae. Milkwort Family.

783. Polygala cruciata L. Crossleaf Milkwort. Gallia, Lucas.
784. Polygala verticillata L. Whorled Milkwort. General; no specimens from the western part of the state.
785. Polygala ambigua Nutt. Loose-spiked Milkwort. Rather general.
786. Polygala viridescens L. Purple Milkwort. General.
787. Polygala senega L. Seneca Snakeroot. Rather general.
788. Polygala polygama Walt. Racemed Milkwort. Ashtabula, Cuyahoga, Lucas.
789. Polygala pauciflora Willd. Fringed Milkwort. Stark County.

Euphorbiaceae. Spurge Family.

790. Phyllanthus carolinensis Walt. Carolina Phyllanthus. Hamilton, Warren.
791. Croton capitatus Mx. Capitate Croton. Hamilton, Franklin.

792. Croton monanthogynus Mx. Single-fruited Croton. Franklin County.

793. Acalypha virginica L. Virginia Three-seeded Mercury. General.

794. Acalypha gracilens Gr. Slender Three-seeded Mercury. Rather general.

795. Acalypha ostryaefolia Ridd. Hornbeam Three-seeded Mercury. Washington County.

796. Mercurialis annua L. Mercury. Lake County. Introduced.

797. Ricinus communis L. Castor-oil-plant. Erie, Ottawa, Franklin. Escaped.

798. Poinsettia dentata (Mx.) Small. Toothed Spurge. Lake, Cuyahoga, Ottawa, Lucas, Franklin, Greene, Warren, Hamilton.

799. Tithymalus lathyrus (L.) Hill. Caper Spurge. Hocking County. From Europe.

800. Tithymalus obtusatus (Pursh) K. & G. Bluntleaf Spurge. Lucas, Wyandot, Auglaize, Jackson, Hamilton.

801. Tithymalus platyphyllus (L.) Hill. Broadleaf Spurge. Ashtabula, Cuyahoga. From Europe.

802. Tithymalus helioscopia (L.) Hill. Sun Spurge. Lake County. From Europe.

803. Tithymalus cyparissias (L.) Hill. Cypress Spurge. General. From Europe.

804. Tithymalus peplus (L.) Hill. Petty Spurge. Clark, Erie, Summit, Lake. From Europe.

805. Tithymalus commutatus (Eng.) K. & G. Tinted Spurge. General.

806. Tithymalus corollata (L.) K. & G. Flowering Spurge. General and abundant.

807. Dichrophyllum marginatum (Pursh) K. & G. Snow-on-the-mountain. Lake, Cuyahoga, Erie, Auglaize, Franklin, Clark, Montgomery, Hamilton. From the West.

808. Chamaesyce preslii (Guss.) Arth. Nodding Spurge. General and abundant.

809. Chamaesyce rafinesqui (Greene) Small. Hairy Spurge. Defiance, Vinton.

810. Chamaesyce humistrata (Eng.) Small. Hairy Spreading Spurge. Cuyahoga, Erie, Belmont, Champaign.

811. Chamaesyce maculata (L.) Small. Spotted Spurge. Rather general.

812. Chamaesyce polygonifolia (L.) Small. Knotweed Spurge. Erie, Cuyahoga, Lake.

813. Chamaesyce serpens. (H. B. K.) Small. Roundleaf Spurge. Ottawa County.

Callitrichaceae. Water-starwort Family.

814. Callitriche austini Eng. Terrestrial Water-starwort. Cuyahoga, Clermont.

815. Callitriche palustris L. Vernal Water-starwort. Auglaize, Lorain, Trumbull.

816. Callitriche heterophylla Pursh. Larger Water-starwort. Cuyahoga County.

Order, *Malvales.*

Malvaceae. Mallow Family.

817. Malva sylvestris L. High Mallow. Cuyahoga, Auglaize. From Europe.

818. Malva rotundifolia L. Roundleaf Mallow. General and abundant. From Europe.

819. Malva verticillata L. Curled Mallow. No specimens. From Europe.

820. Malva alcea L. European Mallow. Escaped in Cuyahoga County. From Europe.

821. Malva moschata L. Musk Mallow. Northern part of the state, as far south as Muskingum County. From Europe.

822. Callirhoe involucrata (T. & G.) Gr. Purple Poppy-mallow. A waif in Franklin County.

823. Althaea officinalis L. Marsh-mallow. A waif in Scioto County.

824. Althaea rosea L. Hollyhock. Lucas, Erie, Madison, Brown, Montgomery, Scioto. Escaped from cultivation.

825. Sida spinosa L. Prickly Sida. Rather general. From the tropics.

826. Sida hermaphrodita (L.) Rusby. Tall Sida. No specimens.

827. Napaea dioica L. Glade-mallow. Defiance, Clark, Madison, Franklin, Fairfield, Highland.

828. Abutilon abutilon (L.) Rusby. Velvet-leaf. General and abundant. From Asia.

829. Hibiscus moscheutos L. Swamp Rose-mallow. Ashtabula, Cuyahoga, Erie, Wayne, Licking, Perry.

830. Hibiscus militaris Cav. Halberd-leaf Rose-mallow. Lucas, Paulding, Auglaize, Defiance, Shelby, Franklin.
831. Hibiscus trionum L. Bladder Ketmia. General. From Europe.

Tiliaceae. Linden Family.

832. Tilia americana L. American Linden. General and abundant.
833. Tilia heterophylla Vent. White Linden. Hamilton, Scioto.
834. Tilia michauxii Nutt. Michaux's Linden. No specimens.

Order, *Violales.*

Hypericaceae. St. John's-wort Family.

835. Hypericum ascyron L. Great St. John's-wort. Rather general.
836. Hypericum kalmianum L. Kalm's St. John's-wort. Ottawa, Erie, Summit.
837. Hypericum prolificum L. Shrubby St. John's-wort. Rather general.
838. Hypericum perforatum L. Common St. John's-wort. General. From Europe.
839. Hypericum punctatum Lam. Spotted St. John's-wort. General.
840. Hypericum cistifolium Lam. Round-podded St. John's-wort. Montgomery, Clermont, Franklin.
841. Hypericum ellipticum Hook. Elliptic-leaf St. John's-wort. Lake County.
842. Hypericum virgatum Lam. Virgate St. John's-wort. Jackson County.
843. Hypericum boreale (Britt.) Bickn. Northern St. John's-wort. Geauga, Defiance, Wayne.
844. Hypericum mutilum L. Small-flowered St. John's-wort. General.
845. Hypericum gymnanthum Eng. & Gr. Clasping-leaf St. John's-wort. Erie, Ottawa.
846. Hypericum majus (Gr.) Britt. Large Canadian St. John's-wort. (Erie County—Moseley Herbarium.)
847. Hypericum canadense L. Canadian St. John's-wort. (Erie County—Moseley Herbarium.)
848 Hypericum drummondii (Grev. & Hook.) T. & G. Drummond's St. John's-wort. Hamilton, Clermont, Ashtabula, Hocking.
849. Sarothra gentianoides L. Orange-grass. Erie, Gallia, Scioto.
850. Triadenum virginicum (L.) Raf. Marsh St. John's-wort. Cuyahoga, Geauga, Erie, Huron, Wayne, Licking.

851. Ascyrum hypericoides L. St. Andrew's-cross. Hamilton, Scioto, Lawrence, Gallia, Jackson, Hocking, Fairfield.

Cistaceae. Rock-rose Family.

852. Crocanthemum majus (L.) Britt. Hoary Frostweed. Lucas, Portage, Fairfield.

853. Crocanthemum canadense (L.) Mx. Canada Frostweed. Erie, Lucas, Wood.

854. Lechea minor L. Thyme-leaf Pinweed. Jefferson, Hocking, Perry.

855. Lechea racemulosa Mx. Oblong-fruited Pinweed. Adams, Scioto, Jackson, Hocking, Fairfield, Licking, Lucas.

856. Lechea villosa Ell. Hairy Pinweed. Summit, Wayne, Erie, Lucas.

857. Lechea tenuifolia Mx. Narrow-leaf Pinweed. Lucas.

858. Lechea leggettii Britt & Holl. Leggett's Pinweed. No specimens.

859. Lechea stricta Legg. Prairie Pinweed. Portage County.

Violaceae. Violet Family.

860. Cubelium concolor (Forst.) Raf. Green Violet. Southern half of state to Auglaize, Licking, and Belmont Counties; also in Lake County.

861. Viola canadensis L. Canada Violet. Southeastern half of state; from Huron to Fairfield and Hamilton Counties.

862. Viola eriocarpa Schw. Smooth Yellow Violet. General.

863. Viola pubescens Ait. Hairy Yellow Violet. General.

864. Viola hastata Mx. Halberd-leaf Yellow Violet. Cuyahoga, Lake, Portage, Columbiana, Belmont.

865. Viola striata Ait. Striped Violet. General and abundant.

866. Viola conspersa Reich. American Dog Violet. Lucas, Wyandot, Lorain, Portage, Trumbull.

867. Viola rostrata Pursh. Long-spurred Violet. Rather general.

868. Viola rafinesquii Greene. Wild Pansy. Hamilton, Montgomery, Miami, Pike, Ross, Franklin, Ottawa, Erie, Lake.

869. Viola tricolor L. Garden Pansy. Cuyahoga, County. From Europe.

870. Viola odorata L. Sweet Violet. Lake, Franklin. From Europe.

871. Viola rotundifolia Mx. Roundleaf Violet. Ashtabula, Cuyahoga, Tuscarawas, Hocking.

872. Viola blanda Willd. Sweet White Violet. Rather general.
873. Viola pallens (Banks) Brain. Woodland White Violet. Cuyahoga, Hancock, Fairfield, Vinton.
874. Viola lanceolata L. Lanceleaf Violet. Lake, Fairfield.
875. Viola affinis Le C. Thinleaf Blue Violet. General and abundant.
876. Viola papilonacea Pursh. Common Blue Violet. General and abundant.
877. Viola hirsutula Brain. Southern Wood Violet. Hocking, Fairfield.
878. Viola sororia Willd. Woolly blue Violet. Lake, Portage, Belmont, Wood Warren,
879. Viola palmata L. Early Blue Violet. General and abundant.
880. Viola triloba Schw. Three-lobed Violet. Lake County.
881. Viola emarginata (Nutt.) Le Conte. Triangle-leaf Violet. Lake County.
882. Viola pedatifida Don. Larkspur Violet. Ottawa, Auglaize.
883. Viola fimbriatula Sm. Ovate-leaf Violet. Wood, Licking, Jefferson, Wayne, Portage, Cuyahoga, Lake.
884. Viola sagittata Ait. Arrowleaf Violet. Franklin, Fulton, Erie, Lorain, Cuyahoga.
885. Viola pedata L. Birdfoot Violet. Lawrence, Scioto.

Passifloraceae. Passionflower Family.

886. Passiflora lutea L. Yellow Passion-flower. Southern half of state to Darke and Franklin Counties.

Subclass, CENTROSPERMAE.

Order, *Caryophyllales.*

Caryophyllaceae. Pink Family.

Subfamily, ALSINATAE.

887. Sagina procumbens L. Procumbent Pearlwort. Lake, Gallia. From Europe.
888. Sagina decumbens (Ell.) T. & G. Decumbent Pearlwort. Lawrence County.
889. Arenaria serpyllifolia L. Thyme-leaf Sandwort. General. From Europe.
890. Arenaria stricta Mx. Rock Sandwort. Ottawa, Erie, Cuyahoga, Clark, Franklin.
891. Arenaria patula Mx. Pitcher's Sandwort. Montgomery County.

892. Moehringia lateriflora (L.) Fenzl. Bluntleaf Moeringia. Ottawa, Auglaize, Morrow, Perry, Franklin, Darke.

893. Holosteum umbellatum L. Jagged Chickweed. Hamilton County. From Europe.

894. Alsine aquatica (L.) Britt. Water Chickweed. Guernsey County. From Europe.

895. Alsine media L. Common Chickweed. General and abundant. From Europe.

896. Alsine pubera (Mx.) Britt. Great Chickweed. Southern Ohio as far north as Preble and Fairfield Counties.

897. Alsine longifolia (Muhl.) Britt. Longleaf Stichwort. General.

898. Alsine graminea (L.) Britt. Lesser Stichwort. Cuyahoga, Auglaize, Belmont. From Europe.

899. Cerastium vulgatum L. Common Mouse-ear Chickweed. General. From Europe.

900. Cerastium longipedunculatum Muhl. Nodding Chickweed. General.

901. Cerastium arvense L. Field Chickweed. Monroe, Trumbull, Ottawa, Sandusky, Miami.

901a. Cerastium arvense webbii Jennings. Cuyahoga County.

902. Cerastium velutinum Raf. Barren Chickweed. Erie, Monroe.

903. Spergula arvensis L. Corn Spurry. Lake County. From Europe.

904. Tissa rubra (L.) Britt. Sand Spurry. Lake County. From Europe.

Subfamily, CARYOPHYLLATAE.

905. Agrostemma githago L. Corn Cockle. General. From Europe.

906. Lychnis coronaria (L.) Desv. Mullen Pink. Lake, Cuyahoga, Portage, Fairfield. From Europe.

907. Lychnis viscaria L. Viscid Lychnis. Lake County. Escaped.

908. Lychnis alba Mill. White Lychnis. Lake, Wayne, Noble, Meigs. From Europe.

909. Lychnis dioica L. Red Lychnis. From Europe. (Erie County—Moseley Herbarium.)

910. Silene stellata (L.) Ait. Starry Campion. General.

911. Silene alba Muhl. White Campion. Butler, Clermont.

912. Silene latifolia (Mill.) Britt & Rend. Bladder Campion. Erie County. From Europe.

913. Silene virginica L. Fire Pink. General.
914. Silene rotundifolia Nutt. Roundleaf Catchfly. Hocking, Jackson.
915. Silene armeria L. Sweet William Catchfly. Monroe, Licking, Cuyahoga, Lake. From Europe.
916. Silene noctiflora L. Night-blooming Catchfly. Greene, Auglaize, Lucas, Sandusky, Erie, Cuyahoga, Lake, Belmont, Jefferson. From Europe.
917. Silene dichotoma Ehrh. Forked Catchfly. Noble County, (Ottawa County—Moseley Herbarium). From Europe.
918. Silene conica L. Striate Catchfly. Sandusky County. From Europe.
919. Silene regia Sims. Royal Catchfly. Clark, Madison.
920. Silene caroliniana Walt. Carolina Catchfly. Washington, Monroe, Jefferson.
921. Silene antirrhina L. Sleepy Catchfly. General.
922. Saponaria officinalis L. Bouncing-Bet. General. From Europe.
923. Vaccaria vaccaria (L.) Britt. Cowherb. Ashtabula, Lake. From Europe.
924. Dianthus prolifer L. Proliferous Pink. Cuyahoga County. From Europe.
925. Dianthus armeria L. Deptford Pink. Jefferson, Gallia, Licking. From Europe.
926. Dianthus deltoides L. Maiden Pink. Lake County. Escaped.
927. Dianthus barbatus L. Sweet William. Portage County. From Europe.

Aizoaceae. Carpetweed Family.

928. Mollugo verticillata L. Carpetweed. General. From the South.

Portulacaceae. Purslane Family.

929. Claytonia virginica L. Spring-beauty. General and abundant.
930. Limnia perfoliata (Donn.) Haw. Spanish-lettuce. No specimens. From the West.
931. Portulaca oleracea L. Purslane. Hamilton, Fayette, Auglaize, Holmes, Huron, Erie, Franklin. From the Southwest.
932. Portulaca grandiflora Hook. Garden Portulaca. Auglaize, Franklin. Escaped.

Nyctaginaceae. Four-o'clock Family.

933. Allionia nyctaginea Mx. Heartleaf Umbrella-wort. Hamilton, Montgomery, Greene, Champaign, Monroe, Erie. From the West.
934. Allionia hirsuta Pursh. Hairy Umbrella-wort. Ashtabula County. From the West.
935. Mirabilis jalapa L. Four-o'clock. Persistent in Franklin County.

Phytolaccaceae. Pokeweed Family.

936. Phytolacca americana L. Pokeweed. General and abundant.

Order, *Chenopodiales.*

Corrigiolaceae. Whitlow-wort Family.

937. Anychia polygonoides Raf. Forked Anychia. Southern part of state as far north as Franklin, Holmes, and Belmont Counties; also in Ottawa and Erie Counties.
938. Anychia canadensis (L.) B. S. P. Slender Anychia General.
939. Scleranthus annuus L. Knawel. Lake County. From Europe.

Amaranthaceae. Amaranth Family.

940. Celosia cristata L. Cock's-comb. Franklin County. Persistent.
941. Amaranthus retroflexus L. Rough Pigweed. General and abundant. From tropical America.
942. Amaranthus hybridus L. Slender Pigweed. General. From tropical America.
943. Amaranthus spinosus L. Spiny Amaranth. Southern half of state. From tropical America.
944. * Amaranthus graecizans L. Tumble-weed (Amaranth). Western half of state, from Hamilton to Franklin and Lake Counties.
945. Amaranthus blitoides Wats. Mat Amaranth. General. From the West.
946. Acnida tuberculata Moq. Tubercled Water-hemp. Rather general.
947. Acnida tamariscina (Nutt.) Wood. Western Water-hemp. Franklin, Clinton. From the West.
948. Iresine paniculata (L.) Ktz. Bloodleaf. No specimens.
949. Gomphrena globosa L. Globe-amaranth. Wood County. Escaped from gardens.

Chenopodiaceae. Goosefoot Family.

950. Chenopodium album L. Lamb's-quarter. General and abundant. From Europe.
951. Chenopodium glaucum L. Oakleaf Goosefoot. Summit, Erie, Ottawa, Lucas, Auglaize. From Europe.
952. Chenopodium leptophyllum (Moq.) Nutt. Narrowleaf Goosefoot. Lake, Lorain.
953. Chenopodium vulvaria L. Fetid Goosefoot. Lorain County. From Europe.
954. Chenopodium polyspermum L. Many-seeded Goosefoot. Lake County. From Europe.
955. Chenopodium boscianum Moq. Bosc's Goosefoot. Washington, Meigs, Ross, Franklin, Miami, Ottawa, Erie.
956. Chenopodium murale L. Nettle-leaf Goosefoot. Rather general. From Europe.
957. Chenopodium hybridum L. Maple-leaf Goosefoot. Rather general.
958. Chenopodium botrys L. Feather Geranium. Rather general. From Europe.
959. Chenopodium ambrosioides L. Mexican Tea. Rather general. From tropical America.
960. Blitum capitatum L. Strawberry Blite. Summit County.
961. Spinacia oleracea L. Common Spinach. Lorain County. Escaped from gardens.
962. Cycloloma atriplicifolium (Spreng.) Coult. Tumbleweed. Erie, Lake. From the West.
963. Kochia scoparia (L.) Roth. Mock-cypress. Morgan, Franklin, Cuyahoga. From Europe.
964. Atriplex hastata L. Halberd-leaf Orache. General.
965. Atriplex rosea L. Red Orache. Erie County. From Europe.
966. Salsola pestifer Nels. Russian-thistle. Williams, Lucas, Auglaize, Ottawa, Erie, Wayne, Cuyahoga, Lake, Portage. From Europe.

Order, *Polygonales*.

Polygonaceae. Buckwheat Family.

967. Rumex altissimus Wood. Tall Dock. Rather general.
968. Rumex verticillatus L. Swamp Dock. Rather general.
969. Rumex mexicanus Meisn. Willow-leaf Dock. Allen County.

970. Rumex patientia L. Patience Dock. From Europe. No specimens.

971. Rumex britannica L. Great Water Dock. Williams, Erie, Coshocton, Licking, Franklin, Hocking, Champaign.

972. Rumex crispus L. Curled Dock. Naturalized from Europe. General and abundant.

973. Rumex conglomeratus Murr. Clustered Dock. Lake County. From Europe.

974. Rumex obtusifolius L. Broadleaf Dock. Naturalized from Europe. General and abundant.

975. Rumex acetosella L. Sheep Sorrel. General and abundant. From Europe.

976. Pleuropterus zuccarinii Small. Japanese Knotweed. Cuyahoga County. From Japan.

977. Tiniaria convolvulus (L.) W. & M. Black Bindweed. General. Naturalized from Europe.

978. Tiniaria scandens (L.) Small. Climbing False Buckwheat. General.

979. Tiniaria dumetorum (L.) Opiz. Copse False Buckwheat. From Europe. (Erie, Ottawa—Moseley Herbarium.)

980. Tracaulon arifolium (L.) Raf. Halberd-leaf Tear-thumb. Northern part of state as far south as Auglaize and Belmont Counties.

981. Tracaulon sagittatum (L.) Small. Arrow-leaf Tear-thumb. General.

982. Fagopyrum fagopyrum (L.) Karst. Buckwheat. From Europe. General.

983. Persicaria amphibia (L.) S. F. Gr. Water Persicaria. Summit, Fairfield, Ottawa, Stark, Franklin, Clark.

984. Persicaria muhlenbergii (Wats.) Small. Swamp Persicaria. Northern half of state to Shelby and Perry Counties.

985. Persicaria lapathifolia (L.) S. F. Gray. Pale Persicaria. General. Naturalized from Europe.

986. Persicaria pennsylvanica (L.) Small. Pennsylvania Persicaria. General and abundant.

987. Persicaria careyi (Olney) Greene. Carey's Persicaria. Erie County.

988. Persicaria persicaria (L.) Small. Lady's-thumb. General and abundant. Naturalized from Europe.

989. Persicaria hydropiperoides (Mx.) Small. Mild Smartweed. General.

990. Persicaria hydropiper (L.) Opiz. Water Smartweed. General. From Europe.

991. Persicaria punctata (Ell.) Small. Dotted Smartweed. General.

992. Persicaria orientalis (L.) Spach. Prince's-feather. Lucas, Montgomery, Franklin, Hocking, Meigs, Scioto. Native of India.

993. Tovara virginiana (L.) Raf. Virginia Knotweed. General.

994. Polygonum aviculare L. Doorweed. General and abundant.

995. Polygonum buxiforme Small. Shore Knotweed. Wayne County.

996. Polygonum erectum L. Erect Knotweed. Lake, Richland, Franklin, Muskingum, Morgan, Hocking, Athens, Meigs, Warren, Clermont.

997. Polygonum ramosissimum Mx. Bushy Knotweed. Lake, Franklin. Waifs from the West.

998. Polygonum tenue Mx. Slender Knotweed. Erie County.

Order, *Piperales.*

Saururaceae. Lizard's-tail Family.

999. Saururus cernuus L. Lizard's-tail. General.

Subclass, CALYCIFLORAE.

Order, *Rosales.*

Rosaceae. Rose Family.

Subfamily, ROSATAE.

1000. Geum rivale L. Purple Avens. Champaign, Geauga.

1001. Geum canaadense Jacq. White Avens. General.

1002. Geum flavum (Port.) Bickn. Cream-colored Avens. No specimens.

1003. Geum virginianum L. Rough Avens. General.

1004. Geum strictum Ait. Yellow Avens. Eastern half of state; also in Preble County.

1005. Geum vernum (Raf.) T. & G. Spring Avens. General.

1006. Dasiphora fruticosa (L.) Rydb. Shrubby Cinquefoil. Rather general.

1007 Potentilla paradoxa Nutt. Bushy Cinquefoil. Erie County.

1008. Potentilla argentea L. Silvery Cinquefoil. Portage, Cuyahoga, Erie, Licking.

1009. Potentilla recta L. Upright Cinquefoil. Lake, Erie, Franklin, Hocking. From Europe.

1010. Potentilla monspeliensis L. Rough Cinquefoil. General.

1011. Potentilla canadensis L. Common Five-finger. General and abundant.

1012. Potentilla pumila Poir. Dwarf Five-finger. Lake, Monroe, Vinton.

1013. Potentilla reptans L. European Five-finger. Lake County. From Europe.

1014. Argentina anserina (L.) Rydb. Silverweed. Cuyahoga, Lorain, Erie, Ottawa, Lucas, Hamilton.

1015. Comarum palustre L. Purple Marshlocks. Ashtabula, Lorain, Portage, Summit, Ashland, Stark, Licking.

1016. Drymocallis agrimonioides (Pursh.) Rydb. Tall Cinquefoil. Erie, Cuyahoga, Lake.

1017. Waldsteinia fragarioides (Mx.) Tratt. Dry Strawberry. Ashtabula, Cuyahoga, Portage, Medina, Clark, Franklin, Greene.

1018. Fragaria americana (Port.) Britt. American Wood Strawberry. Rather general.

1019. Fragaria vesca L. European Wood Strawberry (white-fruited variety). Belmont, Hocking. From Europe.

1020. Fragaria virginiana Duch. Virginia Strawberry. General and abundant.

1021. Rubus frondosus Bigel. Leafy-flowered Blackberry. Lake, Columbiana, Coshocton, Hancock, Gallia.

1022. Rubus alleghaniensis Port. High Blackberry. General and abundant.

1023. Rubus procumbens Muhl. Common Dewberry. General.

1024. Rubus hispidus L. Hispid Dewberry. Ashtabula, Lake, Portage, Summit, Geauga, Cuyahoga, Lucas, Logan.

1025. Rubus occidentalis L. Black Raspberry. General and abundant.

1026. Rubus neglectus Peck. Purple Raspberry. Ashtabula, Stark, Defiance, Williams.

1027. Rubus strigosus Mx. Wild Red Raspberry. Summit, Erie, Clark.

1028. Rubus triflorus Richards. Dwarf Raspberry. Lake, Stark, Lucas, Wood, Sandusky, Wyandot, Champaign, Fairfield, Vinton, Brown.

1029. Rubus phoenicolasius Max. Wineberry. Lake County. Escaped from cultivation.

1030. Rubus odoratus L. Rose-flowered Raspberry. Lake, Cuyahoga, Ashtabula, Summit, Belmont, Jefferson, Monroe, Muskingum, Fairfield.

1031. Porteranthus trifoliatus (L.) Britt. Indian-physic. No specimens.

1032. Porteranthus stipulatus (Muhl.) Britt. American Ipecac. Southern Ohio to Clinton and Guernsey Counties.

1033. Schizonotus sorbifolius (L.) Lindl. Mountain-ash Spiraea. Lake Harrison. From Asia.

1034. Filipendula rubra (Hill.) Rob. Queen-of-the-prairie. Cuyahoga, Erie, Madison, Champaign, Holmes.

1035. Opulaster opulifolius (L.) Ktz. Ninebark. General.

1036. Spiraea alba DuRoi. Narrowleaf Spiraea. General.

1037. Spiraea tomentosa L. Steeple-bush. Eastern half of state west to Cuyahoga, Fairfield, and Jackson Counties; also in Lucas County.

1038. Aruncus aruncus (L.) Karst. Aruncus. Southeastern part of state to Columbiana, Licking, and Scioto Counties.

1039. Dalibarda repens L. Dalibarda. Ashtabula County.

1040. Rosa blanda Ait. Smooth Rose. Lake, Lorain, Erie, Williams, Mercer, Clinton, Clermont.

1041. Rosa carolina L. Swamp Rose. General and abundant.

1042. Rosa virginiana Mill. Virginia Rose. General and abundant.

1043. Rosa rubiginosa L. Swetbrier (Rose). General. From Europe.

1043. Rosa rubiginosa L. Sweetbrier (Rose). General. From Europe.

1044. Rosa gallica L. French Rose. Lake County. Escaped.

1046. Agrimonia parvilflora Sol. Small-flowered Agrimony. General.

1047. Agrimonia gryposepala Wallr. Hairy Agrimony. General.

1048. Agrimonia rostellata Wallr. Woodland Agrimony. Rather general.

1049. Agrimonia striata Mx. Striate Agrimony. Clinton County.

1050. Agrimonia mollis (T. & G.) Britt. Soft Agrimony. Rather general.

1051. Sanguisorba canadensis L. American Burnet. Lake, Cuyahoga, Stark, Miami, Champaign, Clark, Franklin.

1052. Poterium sanguisorba L. Garden Burnet. Lake County. From Europe.

Subfamily, MALATAE.

1053. Sorbus scopulina Greene. Western Mountain-ash. Ashtabula, Ottawa.

1053.1. Sorbus aucuparia L. European Mountain-ash. Lake, Crawford, Escaped.

1054. Pyrus communis L. Pear. Ashtabula, Cuyahog, Summit, Preble, Franklin, Brown. From Europe.

1055. Malus glaucescens Rehd. American Crab-apple. General.

1056. Malus coronaria (L.) Mill. Narrow-leaf Crab-apple. Rather general.

1057. Malus malus (L.) Britt. Common Apple. General. From Europe.

1058. Aronia arbutifolia (L.) Ell. Red Chokeberry. Ashtabula, Geauga, Stark, Licking, Wood.

1059. Aronia atropurpurea Britt. Purple Chokeberry. Licking County.

1060. Aronia melanocarpa (Mx.) Britt. Black Chokeberry. Rather general.

1061. Amelanchier canadensis (L.) Med. Common Juneberry. General.

1062. Amelanchier sanguinea (Pursh.) DC. Roundleaf Juneberry. Highland, Franklin, Erie, Lorain.

1063. Crataegus crus-galli L. Cockspur Hawthorn. General.

1064. Crataegus cuneiformis (Marsh.) Eggl. Marshall's Hawthorn. Rather general.

1065. Crataegus punctata Jacq. Dotted Hawthorn. General.

1066. Crataegus margaretta Ashe. Margaret Hawthorn. From Adams, Franklin, and Lucas Counties westward.

1067. Crataegus succulenta Schrad. Long-spined Hawthorn. Gengeneral.

1068. Crataegus calpodendron (Ehrh.) Medic. Pear Hawthorn. General.

1069. Crataegus brainerdi Sarg. Brainerd's Hawthorn. Lucas, Richland.

1070. Crataegus chrysocarpa Ashe. Roundleaf Hawthorn. Williams County.

1071. Crataegus straminea Beadle. Allegheny Hawthorn. Knox, Franklin, Hocking.

1072. Crataegus boyntoni Beadle. Boynton's Hawthorn. Adams, Morgan, Noble, Guernsey, Tuscarawas.

1073. Crataegus macrosperma Ashe. Large-seeded Hawthorn. General.
1074. Crataegus leiophylla Sarg. Maine's Hawthorn. Jefferson County.
1075. Crataegus beata Sarg. Dunbar's Hawthorn. Lucas, Brown.
1076. Crataegus pruinosa (Wendl.) Koch. Pruinose Hawthorn. General.
1077. Crataegus pringlei Sarg. Pringle's Hawthorn. Williams County.
1078. Crataegus coccinea L. Scarlet Hawthorn. General, but no specimens from the northern counties.
1079. Crataegus albicans Ashe. Tatnall's Hawthorn. Brown, Ross, Jefferson, Ashtabula, Lake, Lucas.
1080. Crataegus mollis (T. & G.) Scheele. Downy Hawthorn. General.
1081. Crataegus monogyna Jacq. May Hawthorn. Williams, Lake, Cuyahoga, Medina, Franklin. From Europe.
1082. Crataegus phaenopyrum (L. f.) Medic. Washington Hawthorn. Jefferson County.
1083. Cotoneaster pyracantha (L.) Spach. Fire-thorn. Franklin County. From Europe.

Subfamily, AMYGDALATAE.

1084. Prunus virginiana L. Black Cherry. (Padus). General and abundant.
1085. Prunus nana DuRoi. Choke Cherry. (Padus). Rather general.
1086. Prunus mahaleb L. Mahaleb Cherry. Lake, Franklin. From Europe.
1087. Prunus pennsylvanica L. f. Red Cherry. Cuyahoga County.
1088. Prunus avium L. Sweet Cherry. Ashtabula, Cuyahoga, Erie, Ottawa, Summit, Ross. From Europe.
1089. Prunus cerasus L. Sour Cherry. Cuyahoga, Summit, Jefferson, Gallia, Clinton. From Europe.
1090. Prunus cuneata Raf. Appalachian Cherry. No specimens.
1091. Prunus pumila L. Sand Cherry. Erie County.
1092. Prunus americana Marsh. Wild Plum. General and abundant.
1093. Amygdalus persica L. Peach. Rather general. Native of Asia.

Fabaceae. Bean Family.

Subfamily, MIMOSATAE.

1094. Acuan illinoensis (Mx.) Ktz. Illinois Acuan. Hamilton, Clermont, Ashtabula.

Subfamily, CASSIATAE.

1095. Cersis canadensis L. Redbud. Rather general, but no specimen from the northeastern counties except Carroll.

1096. Cassia marylandica L. Wild Senna. General.

1097. Cassia medsgeri Shaf. Medsger's Senna. Stark, Washington, Monroe, Franklin, Ottawa.

1098. Chamaecrista nictitans (L.) Moench. Sensitive-pea. Adams, Butler, Gallia, Scioto, Jackson, Hocking, Fairfield, Licking, Stark.

1099. Chamaecrista fasciculata (Mx.) Greene. Large-flowered Sensitive-pea. Rather general in western half of state; also in Lake County.

1100. Gleditsia triacanthos L. Honey-locust. General.

1101. Gymnocladus dioica (L.) Koch. Coffee-bean. General.

Subfamily, FABATAE.

1102. Baptisia australis (L.) R. Br. Blue Wild-indigo. Hamilton, Meigs, Monroe, Lake.

1103. Baptisia tinctoria (L.) R. Br. Yellow Wild-indigo. Trumbull, Lake, Portage, Cuyahoga, Erie, Lucas, Wood.

1104. Baptisia leucantha T. & G. Large White Wild-indigo. Franklin, Crawford, Erie, Defiance.

1105. Crotalaria sagittalis L. Rattlebox. Franklin County. A waif.

1106. Lupinus perennis L. Wild Lupine. Portage, Erie, Sandusky, Wood, Fulton.

1107. Medicago sativa L. Alfalfa. Rather general. From Europe.

1108. Medicago lupulina L. Hop Medic. General. From Europe.

1109. Medicago hispida Gaertn. Toothed Medic. Lake County. From Europe.

1110. Melilotus alba Desv. White Sweet-clover. General and abundant. From Europe.

1111. Melilotus officinalis (L.) Lam. Yellow Sweet-clover. Rather general. From Europe.

1112. Trifolium agrarium L. Yellow Hop Clover. Ashtabula, Lake, Cuyahoga, Knox, Clermont, From Europe.

1113. Trifolium procumbens L. Low Hop Clover. Lake, Cuyahoga, Ottawa, Franklin, Montgomery, Gallia. From Europe.

1114. Trifolium dubium Sibth. Least Hop Clover. Rather general. From Europe.

1115. Trifolium incarnatum L. Crimson Clover. Rather general. From Europe.

1116. Trifolium arvense L. Rabbit-foot Clover. Warren, Stark, Cuyahoga, Lake. From Europe.

1117. Trifolium pratense L. Red Clover. General and abundant. Naturalized from Europe.

1118. Trifolium reflexum L. Buffalo Clover. No specimens.

1119. Trifolium stolniferum Muhl. Running Buffalo Clover. Hamilton, Clermont, Butler, Clinton, Clark, Franklin.

1120. Trifolium hybridum L. Alsike Clover. General. Introduced from Europe.

1121. Trifolium repens L. White Clover. General and abuntdant. Naturalized from Europe.

1122. Lotus corniculatus L. Bird's-foot Trefoil. Lake County. From Europe.

1123. Hosackia americana (Nutt.) Piper. Prarie Bird's-foot. Trefoil. Franklin county. A waif from the west.

1124. Psoralea stipulata T & G. Large-stipuled Psoralea. No specimen.

1125. Psoralea peduncultata (Mill.) Vail. Long-peduncled Psoralea. Erie, Scioto, Lawrence.

1126. Psoralea onobrychis Nutt. Sainfoin Psoralea. In the southwestern fourth of the state.

1127. Amorpha fruticosa L. False Indigo. Lucas County.

1128. Petalostemum purpureum (Vent.) Rydb. Violet Prairie-clover. A waif in Franklin County.

1129. Cracca virginiana I. Goat's-rue. Fulton, Lucas, Erie, Portage, Fairfield, Hocking, Washington, Jackson, Lawrence, Adams, Hamilton.

1130. Robinia pseudoacacia L. Common Locust. General.

1131. Robinia viscosa Vent. Clammy Locust. Ashtabula, Lake, Cuyahoga, Fairfield.

1132. Astragalus carolinianus L. Carolina Milk-vetch. Rather general.

1133. Phaca neglecta T. & G. Cooper's Milk-vetch. Ashtabula, Cuyahoga, Hamilton.

1134. Coronilla varia L. Coronilla. Lake, Brown. From Europe.

1135. Stylosanthes biflora (L.) B. S. P. Pencil-flower. Adams, Scioto, Lawrence, Gallia, Jackson, Hocking.

1136. Meibomia nudiflora (L.) Ktz. Naked-flowered Tick-trefoil. General.

1137. Meibomia grandiflora (Walt.) Ktz. Pointed-leaf Tick-trefoil. General.

1138. Meibomia pauciflora (Nutt.) Ktz. Few-flowered Tick-trefoil. Clermont, Clinton, Auglaize.

1139. Meibomia michauxii Vail. Prostrate Tick-trefoil. Rather general.

1140. Meibomia sessilifolia Torr.) Ktz. Sessile-leaf. Tick-trefoil. Erie, Wood.

1141. Meibomia canescens (L.) Ktz. Hoary Tick-trefoil. General.

1142. Meibomia bracteosa (Mx.) Ktz. Large-bracted Tick-trefoil. Rather general.

1143. Meibomia paniculata (L.) Ktz. Panicled Tick-trefoil. General.

1144. Meibomia viridiflora (L.) Ktz. Velvet-leaf Tick-trefoil. Gallia, Hocking, Cuyahoga.

1145. Meibomia dillenii (Darl.) Ktz. Dillen's Tick-trefoil. General.

1146. Meibomia illinoensis (Gr.) Ktz. Illinois Tick-trefoil. Erie, Ottawa,

1147. Meibomia canadensis (L.) Ktz. Canadian Tick-trefoil. Montgomery, Clark, Auglaize, Fulton, Wood, Erie, Cuyahoga.

1148. Meibomia rigida (Ell.) Ktz. Rigid Tick-trefoil. Paulding, Fairfield.

1149. Meibomia marylandica (L.) Ktz. Maryland Tick-trefoil. Hocking, Fairfield.

1150. Meibomia obtusa (Muhl.) Vail. Ciliate Tick-trefoil. Summit, Erie, Licking.

1151. Lespedeza repens (L.) Bart. Creeping Bush-clover. From Hocking, Franklin and Madison Counties southward; also in Cuyahoga County.

1152. Lespedeza procumbens Mx. Trailing Bush-clover. Fairfield, Wayne, Greene.

1153. Lespedeza nuttallii Darl. Nuttall's Bush-clover. (Moseley herbarium—Erie County.)

1154. Lespedeza violacea (L.) Pers. Violet Bush-clover. Rather general but no specimen from the northeastern counties.

1155. Lespedeza stuvei Nutt. Stuve's Bush-clover. No specimens. (Moseley Herbarium—Erie County).

1156. Lespedeza frutescens (L.) Britt. Wand-like Bush-clover. Rather general.

1157. Lespedeza virginica (L.) Britt. Slender Bush-clover. Scioto, Franklin, Erie.

1158. Lespedeza simulata Mack & Bush. Intermediate Bush-clover. No specimens.

1159. Lespedeza hirta (L.) Horn. Hairy Bush-clover. Rather general.

1160. Lespedeza capitata Mx. Round-headed Bush-clover. Defiance, Fulton, Wood, Ottawa, Erie, Cuyahoga, Franklin, Madison, Highland, Fairfield.

1161. Vicia cracca L. Cow Vetch. Cuyahoga, Lake, Columbiana, Wayne, Huron, Seneca.

1162. Vicia americana Muhl. American Vetch. Cuyahoga, Geauga, Erie, Ottawa, Lucas.

1163. Vicia caroliniana Walt. Carolina Vetch. In the northern and southern counties; also in Darke County.

1164. Vicia tetrasperma (L.) Moench. Slender Vetch. Lake County. From Europe.

1165. Vicia hirsuta (L.) Koch. Hairy Vetch. Lake, Sandusky, Knox, From Europe.

1166. Vicia sativa L. Common Vetch. Ottawa, Hamilton. From Europe.

1167. Vicia angustifolia L. Narrow-leaf Vetch. Lake County. From Europe.

1168. Lathyrus maritimus (L.) Bigel. Beach Pea. Ashtabula, Lake, Cuyahoga, Erie.

1169. Lathyrus venosus Muhl. Veiny Pea. Erie, Williams.

1170. Lathyrus palustris L. Marsh Pea. Lake, Cuyahoga, Summit, Wayne, Erie, Madison, Greene.

1171. Lathyrus myrtifolius Muhl. Myrtle-leaf Marsh Pea. Lake, Cuyahoga, Stark, Erie, Lucas, Defiance, Auglaize.

1172. Lathyrus ochroleucus Hook. Cream-colored Pea. Lake, Cuyahoga, Lorain, Ottawa.

1173. Lathyrus pratensis L. Meadow Pea. Lake County. From Europe.

1174. Dolichos lablab L. Hyacinth Bean. Franklin County. Escaped from gardens.

1175. Glycine apios L. Ground-nut. Rather general.

1176. Falcata comosa (L.) Ktz. Hog-peanut. General.

1177. Falcata pitcheri (T. & G.) Ktz. Pitcher's Hog-peanut. Rather general.

1178. Phaseolus polystachyus (L.) B. S. P. Wild Bean. No specimens.

1179. Phaseolus nanus L. Bush Bean. Auglaize County. Introduced.

1180. Strophostyles helvola (L.) Britt. Trailing Wild Bean. In the lake shore counties and from Hocking County southward; also in Tuscarawas County.

Order, *Saxifragales.*

Crassulaceae. Orpine Family.

Subfamily, CRASSULATAE.

1181. Sedum triphyllum (Haw.) S. F. Gr. Live-forever. Williams, Erie, Knox, Coshocton, Franklin. From Europe.

1182. Sedum telephioides Mx. American Orpine. Adams County.

1183. Sedum acre L. Wall-pepper. Franklin, Ottawa. From Europe.

1184. Sedum ternatum Mx. Wild Stonecrop. General.

Subfamily, PENTHORATAE.

1185. Penthorum sedoides L. Ditch Stonecrop. General and abundant.

Podostemaceae. River-weed Family.

1186. Podostemon ceratophyllum Mx. River-weed. No specimens.

Saxifragaceae. Saxifrage Family.

1187. Micranthes pennsylvanica (L.) Haw. Pennsylvania Saxifrage. Fulton, Auglaize, Clark, Richland, Lorain, Geauga, Stark.

1188. Micranthes virginiensis (Mx.) Small. Early Saxifrage. Eastern half of state to Cuyahoga and Ross Counties; also in Hamilton County.

1189. Sullivantia sullivantii (T. & G.) Britt. Sullivantia. Adams, Highland, Hocking.

1190. Tiarella cordifolia L. False Mitrewort. Cuyahoga, Lorain, Huron, Belmont, Gallia, Highland.

1191. Heuchera americana L. Alum-root. General and abundant.

1192. Mitella diphylla L. Two-leaf Bishop's-cap. General.

1193. Chrysosplenium americanum Schw. Golden Saxifrage. Cuyahoga, Belmont, Stark, Fairfield.

Order, *Thymeleales.*

Lythraceae. Loosestrife Family.

1194. Ammannia coccinea Rottb. Longleaf Ammannia. Erie County.
1195. Rotala ramosior (L.) Koehne. Rotala. Hamilton, Licking, Ottawa.
1196. Decodon verticillatus (L.) Ell. Swamp Loosestrife. Rather general.
1197. Lythrum alatum Pursh. Wing-angled Loosestrife. Rather general; no specimens from the southeast.
1198. Lythrum salicaria L. Spiked Loosestrife. Lake, Cuyahoga. From Europe.
1199. Parsonsia petiolata (L.) Rusby. Blue Waxweed. Southern half of state; also in Cuyahoga and Wayne Counties.

Melastomaceae. Meadow-beauty Family.

1200. Rhexia virginica L. Virginia Meadow-beauty. Erie County.

Thymeleaceae. Mezereum Family.

1201. Dirca palustris L. Leatherwood. Rather general.

Elaeagnaceae. Oleaster Family.

1202. Lepargyraea canadensis (L.) Greene. Canadian Buffalo-berry. Erie, Cuyahoga, Lake.

Order, *Celastrales.*

Rhamnaceae. Buckthorn Family.

1203. Rhamnus cathartica L. Common Buckthorn. Greene County. From Europe.
1204. Rhamnus lanceolata Pursh. Lanceleaf Buckthorn. From Delaware County southward and southwestward.
1205. Rhamnus alnifolia L'Her. Alderleaf Buckthorn. Lake, Cuyahoga, Stark, Champaign.
1206. Rhamnus caroliniana Walt. Carolina Buckthorn. Adams County.
1207 Ceanothus americanus L. Common New Jersey Tea. General in eastern part of the state westward to Ottawa, Clark, Greene, and Adams Counties.
1208. Ceanothus ovatus Desf. Smaller New Jersey Tea. Erie, Ottawa, Crawford.

Vitaceae. Grape Family.

1209. Vitis labrusca L. Northern Fox Grape. Rather general.
1210. Vitis aestivalis Mx. Summer Grape. General.
1211. Vitis bicolor LeC. Winter Grape. General in eastern part of state to Knox and Adams Counties; also in Williams County.
1212. Vitis vulpina L. Riverside Grape. General and abundant.
1213. Vitis cordifolia Mx. Frost Grape. Rather general.
1214. Ampelopsis cordata Mx. Heartleaf Ampelopsis. Scioto County.
1215. Parthenocissus quinquefolia (L.) Planch. Virginia Creeper. General and abundant.

Celastraceae. Stafftree Family.

1216. Euonymus atropurpureus Jacq. Wahoo. General and abundant.
1217. Euonymus europaeus L. Spindletree. Lake County. Escaped.
1218. Euonymus obovatus Nutt. Running Strawberry-bush. Rather general.
1219. Euonymus americanus L. American Strawberry-bush. No specimens.
1220. Celastrus scandens L. Waxwork. General.

Ilicaceae. Holly Family.

1221. Nemopanthus mucronata (L.) Trel. Mountain Holly. Stark, Summit, Defiance, Williams.
1222. Ilex verticillata (L.) Gr. Winterberry. General.
1223. Ilex opaca Ait. American Holly. Lawrence County.

Staphyleaceae. Bladdernut Family.

1224. Staphylea trifolia L. American Bladdernut. General and abundant.

Order, *Sapindales.*

Sapindaceae. Soap-berry Family.

1225. Cardiospermum halicacabum L. Balloon-vine. No specimens. Native of tropical America.

Aesculaceae. Buckeye Family.

1226. Aesculus hippocastanum L. No specimens. Native of Asia.

1227. Aesculus glabra Willd. Ohio Buckeye. General and abundant.
1228. Aesculus octandra Marsh. Yellow Buckeye. Southern part of the state, north to Monroe and Fairfield Counties.

Aceraceae. Maple Family

1229. Acer spicatum Lam. Mountain Maple. From Lorain, Wayne, and Muskingum Counties eastward; also in Greene County
1230. Acer saccharum Marsh. Sugar Maple. General and abundant.
1231. Acer nigrum Mx. Black Maple. General and abundant.
1232. Acer rubrum L. Red Maple. General.
1233. Acer saccharinum L. Silver Maple. General and abundant.
1234. Acer negundo L. Boxelder. General and abundant.

Anacardiaceae. Sumac Family.

1235. Rhus copallina L. Mountain Sumac. Rather general, but no specimens from the west central part of the state.
1236. Rhus hirta (L.) Sudw. Staghorn Sumac. Rather general.
1237. Rhus glabra L. Smooth Sumac. General and abundant.
1238. Schmaltzia crenata (Mill.) Greene. Fragrant Sumac. Western two-thirds of the state.
1239. Toxicodendron vernix (L.) Ktz. Poison Sumac. Geauga, Cuyahoga, Lorain, Wayne, Wyandot, Licking, Fairfield.
1240. Toxicodendron radicans (L.) Ktz. Poison Ivy. General and abundant.
1241. Cotinus cotinus (L.) Sarg. European Smoketree. Escaped in Jefferson County.

Subclass AMENTIFERAE.

Order, *Platanales.*

Hamamelidaceae. Witch-hazel Family.

Subfamily, HAMAMELIDATAE.

1242. Hamamelis virginiana L. Witch-hazel. General, but no specimens from the west central counties.

Subfamily, ALTINGIATAE.

1243. Liquidambar styraciflua L. Sweet-gum. Gallia, Lawrence, Scioto, Adams, Brown, Greene.

Platanaceae. Plane-tree Family.

1244. Platanus occidentalis L. Sycamore. General and abundant.

Order, *Urticales.*

Ulmaceae. Elm Family.

1245. Ulmus americana L. White Elm. General and abundant.
1246. Ulmus thomasi Sarg. Cork Elm. Cuyahoga, Lorain, Ottawa, Huron, Hancock, Hardin, Logan, Franklin, Lawrence, Greene.
1247. Ulmus fulva Mx. Slippery Elm. General and abundant.
1248. Celtis occidentalis L. Common Hackberry. (Including C. crassifolia—young plants and vigorous shoots.) General and abundant, but no specimens from the extreme northeastern counties.

Moraceae. Mulberry Family.

Subfamily, MORATAE.

1249. Morus rubra L. Red Mulberry. General.
1250. Morus alba L. White Mulberry. Introduced. Erie, Lorain, Medina, Summit, Carroll, Montgomery, Clermont, Lawrence.
1251. Toxylon pomiferum Raf. Osage-orange. From the Southwest. General except in the northeastern counties.

Subfamily, CANNABINATAE.

1252. Humulus lupulus L. Hop. General. Introduced.
1253. Humulus japonicus S. & Z. Japanese Hop. A waif in Lucas County.
1254. Cannabis sativa L. Hemp. Clermont, Greene, Franklin, Belmont, Coshocton, Holmes, Lucas. Introduced.

Urticaceae. Nettle Family.

1255. Urtica dioica L. Stinging Nettle. From Europe. Jefferson, Cuyahoga, Lorain.
1256. Urtica gracilis L. Slender Nettle. General.
1257. Urtica urens L. Small Nettle. From Europe. Lake County.
1258. Urticastrum divaricatum (L.) Ktz. Wood Nettle. General.
1259. Pilea pumila (L.) Gr. Clearweed. General.
1260. Boehmeria cylindrica (L.) Sw. False Nettle. General.
1261. Parietaria pennsylvanica Muhl. Pellitory. General.

Order, *Fagales*.

Fagaceae. Beech Family.

1262. Fagus grandifolia Ehrh. American Beech. General and abundant.

1263. Castanea dentata (Marsh.) Borkh. Chestnut. Eastern half of state to Lorain, Franklin, and Adams Counties.

1264. Quercus prinus L. Rock Chestnut Oak. Eastern and southern parts of the state to Cuyahoga, Fairfield, and Clermont Counties.

1265. Quercus muhlenbergii Engelm. Chestnut Oak. General, but no specimens east of Erie nor north of Muskingum and Monroe Counties.

1266. Quercus prinoides Willd. Scrub Chestnut Oak. Starke County.

1267. Quercus bicolor Willd. Swamp White Oak. General.

1268. Quercus alba L. White Oak. General and abundant.

1269. Quercus stellata Wang. Post Oak. From Madison and Morgan Counties southward.

1270. Quercus macrocarpa Mx. Bur Oak. General in the western half of the state to Erie, Franklin, and Fairfield Counties; also in Ashtabula and Belmont Counties.

1271. Quercus imbricaria Mx. Shingle Oak. General.

1271a. Quercus imbricaria X velutina. A hybrid in Hamilton, Licking, Harrison, Erie and Lucas Counties.

1272. Quercus marilandica Muench. Black-Jack (Oak). Adams, Lawrence.

1273. Quercus ilicifolia Wang. Bear Oak. No specimens.

1273.1. Quercus triloba Mx. Spanish Oak. Lawrence County.

1274. Quercus velutina Lam. Quercitron Oak. General.

1275. Quercus coccinea Wang. Scarlet Oak. Hamilton, Auglaize, Fairfield, Franklin, Richland.

1276. Quercus rubra L. Red Oak. General and abundant.

1277. Quercus palustris DuRoi. Pin Oak. General.

Betulaceae. Birch Family.

1278. Carpinus caroliniana Walt. Blue-beech. General and abundant.

1279. Ostyra virginiana (Mill.) Willd. Hop-hornbeam. General and abundant.

1280. Corylus americana Walt. Common Hazelnut. General.

1281. Betula lenta L. Sweet Birch. Fairfield, Hocking, Adams, Scioto.

1282. Betula lutea Mx. f. Yellow Birch. Ashtabula, Stark, Summit, Lake, Cuyahoga, Lorain, Wayne, Fairfield, Hocking.

1283. Betula nigra L. River Birch. From Fairfield and Perry Counties southward.

1284. Betula alba L. European White Birch. Escaped in Lake County.

1285. Betula pumila L. Low Birch. Summit, Stark, Wyandot, Champaign.

1286. Alnus incana (L.) Willd. Hoary Alder. Cuyahoga, Lake, Geauga.

1287. Alnus rugosa (DuRoi), Spreng. Smooth Alder. Eastern half of state, to Lorain, Fairfield and Scioto Counties.

Juglandaceae. Walnut Family.

1288. Hicoria cordiformis (Wang.) Britt. Bitternut (Hickory). General and abundant.

1289 Hicoria microcarpa (Nutt.) Britt. Small Pignut (Hickory). General.

1290 Hicoria glaba (Mill.) Britt. Pignut (Hickory.) Rather general, but no speicmens from the west central part.

1291. Hicoria alba (L.) Britt. Mockernut (Hickory). Rather general, but no specimens from the extreme eastern and extreme western counties.

1292. Hicoria laciniosa (Mx. f.) Sarg. Shellbark (Hickory). Huron, Wyandot, Licking, Franklin, Pickaway, Scioto, Clermont.

1293. Hicoria ovata (Mill.) Britt. Shagbark (Hickory). General and abundant.

1294. Juglans nigra L. Black Walnut. General and abundant.

1295. Juglans cinerea L. Butternut. General, but no specimens from the northeastern counties.

Myricaceae. Bayberry Family.

1296. Comptonia peregrina (L.) Coult. Comptonia. Lake, Erie, Portage, Lucas, Fulton, Wood, Knox.

Order, *Salicales.*

Salicaceae. Willow Family.

1297. Populus alba L. White Poplar. General. Introduced.

1298. Populus heterophyla L. Swamp Poplar. Lake, Huron, Richland, Williams, Auglaize, Shelby, Logan, Knox, Licking.

1299. Populus balsamifera L. Balsam Poplar. Huron, Ashtabula, Geauga, Carroll.

1299a. Populus balsamifera candicans (Ait.) Gr. Balm-of-Gilead. Preble, Clermont, Franklin, Hocking, Coshocton, Harrison, Jefferson, Lorain. Escaped from cultivation.

1300. Populus deltoides Marsh. Cottonwood. General.

1301. Populus italica Moench. Lombardy Poplar. Rather general; probably mostly planted. From Europe.

1302. Populus grandidentata Mx. Largetooth Aspen. General.

1303. Populus tremuloides Mx. American Aspen. General in the northern part of the state, south to Franklin and Hocking Counties, also in Adams County.

1304. Salix amygdaloides And. Peachleaf Willow. Northwestern half of state, from Ashtabula to Franklin Counties westward.

1305. Salix nigra Marsh. Black Willow. General and abundant.

1305a. Salix nigra X amygdaloides. Erie, Ashtabula.

1306. Salix lucida Muhl. Shining Willow. Northern part of state as far south as Logan and Knox Counties.

1307. Salix fragilis L. Crack Willow. General. From Europe.

1307a. Salix fragilis X alba. Franklin and Ottawa Counties.

1308. Salix pentandra L. Bayleaf Willow. Escaped in Franklin County.

1309. Salix alba L. White Willow. General. Native of Europe.

1309a. Salix alba X lucida. Ashtabula, Logan.

1309b. Salix alba X babylonica. Ashtabula County.

1310. Salix babylonica L. Weeping Willow. Native of Asia. Ashtabula, Wayne.

1310a. Salix babylonica X fragilis. Erie County.

1311. Salix interior Rowlee. Sandbar Willow. General and abundant.

1311a. Salix interior wheeleri Rowlee. Preble, Erie, Lake.

1312. Salix glaucophylla Bebb. Broadleaf Willow. Erie, Wyandot.

1313. Salix cordata Muhl. Heartleaf Willow. General.

1313a. Salix cordata X sericea. Ashtabula County.

1314. Salix adenophylla Hook. Furry Willow. Erie County.

1315. Salix candida Fl. Hoary Willow. Wyandot, Erie.

1316. Salix sericea Marsh. Silky Willow. General.

1317. Salix petiolaris Sm. Slender Willow. Erie, Wood, Lucas, Fulton.
1318. Salix bebbiana Sarg. Bebb's Willow. Northern counties from Ashtabula to Fulton County; also in Wyandot County.
1319. Salix discolor Muhl. Pussy Willow. General and abundant.
1320. Salix humilis Marsh. Prairie Willow. Lake, Wood, Lucas, Fulton, Fairfield, Hocking.
1320a. Salix humilis tristis (Ait.) Griggs. Athens, Madison.
1321. Salix purpurea L. Purple Willow. Native of Europe. Rather general.
1322. Salix pedicellaris Pursh. Bog Willow. Williams, Portage, Wayne, Licking, Perry.

Subclass, MYRTIFLORAE.

Order, *Cactales*.

Cactaceae. Cactus Family.

1323. Opuntia humifusa Raf. Western Prickly-pair. Erie, Scioto.

Order, *Myrtales*.

Hydrangeaceae. Hydrangea Family.

Subfamily, PHILADELPHATAE.

1324. Philadelphus coronarius L. Garden Mock-orange. Erie, Auglaize, Belmont, Jefferson, Monroe. From Europe.

Subfamily, HYDRANGEATAE.

1325. Hydrangea arborescens L. Wild Hydrangea. Southern half of state to Champaign and Mahoning Counties.

Grossulariaceae. Gooseberry Family.

1326. Ribes lacustre (Pers.) Poir. Swamp Currant. (Erie County—Moseley Herbarium.)
1327. Ribes vulgare Lam. Red Currant. Fulton, Lorain, Cuyahoga, Ashtabula. Native of Europe.
1328. Ribes americanum Mill. Wild Black Currant. General.
1329. Ribes odoratum Wendl. Buffalo Currant. Belmont, Hocking, Franklin, Auglaize, Richland, Lake. From the west.
1330. Grossularia cynosbati (L.) Mill. Wild Gooseberry. General.
1331. Grossularia oxyacanthoides (L.) Mill. Northern Gooseberry. Stark, Wayne.

1332. Grossularia hirtella (Mx.) Spach. Low Gooseberry. Geauga, Summit, Champaign,

1333. Grossularia reclinata (L.) Mill. Garden Gooseberry. Lawrence, Franklin. From Europe.

Onagraceae. Evening-primrose Family.

1334. Ludwigia polycarpa S. & P. Many-fruited Ludwigia. Cuyahoga, Lucas, Auglaize, Hocking.

1335. Ludwigia alternifolia L. Seed-box. Rather general.

1336. Isnardia palustris L. Marsh Purslane. General.

1337. Chamaenerion angustifolium (L.) Scop. Fire-weed. Northern fourth of the state.

1338. Epilobium lineare Muhl. Linear-leaf. Willow-herb. Portage, Erie, Ottawa, Clark.

1339. Epilobium strictum Muhl. Downy Willow-herb. Licking County.

1340. Epilobium coloratum Muhl. Purple Willow-herb. General.

1341. Epilobium adenocaulon Haussk. Northern Willow-herb. Northern Ohio, south to Franklin and Madison Counties.

1342. Oenothera biennis L. Common Evening-primrose. General.

1343. Oenothera oakesiana Robb. Oakes' Evening-primrose. (Erie County—Moseley Herbarium.)

1344. Raimannia laciniata (Hill.) Rose. Cutleaf Evening-primrose. Cuyahoga County.

1345. Kneiffia pratensis Small. Meadow Sundrops. No specimens.

1346. Kneiffia pumila (L.) Spach. Small Sundrops. Eastern part of Ohio as far west as Cuyahoga and Hocking Counties.

1347. Kneiffia fruticosa (L.) Raim. Common Sundrops. Rather general, but no specimens from the western part of the state.

1348. Hartmannia speciosa (Nutt.) Small. White Evening-primrose. A waif in Franklin County.

1349. Lavauxia triloba (Nutt.) Spach. Three-lobed Evening-primrose. From two localities in Montgomery County.

1350. Gaura biennis L. Biennial Gaura. Rather general.

1351. Circaea lutetiana L. Common Enchanter's-nightshade. General and abundant.

1352. Circaea intermedia Ehrh. Intermediate Enchanter's-nightshade. No specimens.

1353. Circaea alpina L. Small Enchanter's-nightshade. Ashtabula, Cuyahoga, Lorain, Summit, Crawford, Clark, Hocking.

Haloragidaceae. Water-milfoil Family.

1354. Myriophyllum spicatum L. Spiked Water-milfoil. Rather general.

1355. Myriophyllum verticillatum L. Whorled Water-milfoil. Erie County.

1356. Myriophyllum heterophyllum Mx. Variant-leaf Water-milfoil. No specimens.

1357. Proserpinaca palustris L. Mermaid-weed. Ashtabula, Cuyahoga, Portage, Wayne, Erie, Wyandot.

Order, *Loasales.*

Cucurbitaceae. Gourd Family.

1358. Cucurbita pepo L. Pumpkin. Erie, Wood, Spontaneous.

1359. Cucurbita maxima Duchesne. Squash. Brown County. Spontaneous.

1360. Citrullus citrullus (L.) Karst. Watermelon. Athens, Ottawa, Spontaneous.

1361. Cucumis sativus L. Cucumber. No specimens. Sometimes spontaneous.

1362. Cucumis melo L. Muskmelon. Madison, Franklin. Spontaneous.

1363. Micrampelis lobata (Mx.) Greene. Wild Balsam-apple. Rather general.

1364. Sicyos augulatus L. Star Cucumber. Rather general.

Order, *Aristolochiales.*

Aristolochiaceae. Birthwort Family.

1365. Asarum canadense L. Canadian Wild-ginger. General.

1366. Asarum acuminatum (Ashe.) Bicken. Long-tipped Wild-ginger. Rather general.

1367. Asarum reflexum Bicken. Short-lobed Wild-ginger. Rather general.

1368. Aristolochia serpentaria L. Virginia Snakeroot. Rather general.

Order, *Santalales.*

Santalaceae. Sandalwood Family.

1369. Comandra umbellata (L.) Nutt. Bastard Toad-flax. Rather general.

Loranthaceae. Mistletoe Family.

1370. Phoradendron flavescens (Pursh) Nutt. American Mistletoe. Southern counties, as far north as Ross and Athens.

Subclass, HETEROMERAE.

Order, *Primulales.*

Primulaceae. Primrose Family.

1371. Lysimachia vulgaris L. Common Yellow Loosestrife. Lake County. From Europe.
1372. Lysimachia quadrifolia L. Whorled Yellow Loosestrife. General in the eastern half of the state; also in Fulton and Adams Counties.
1373. Lysimachia terrestris (L.) B. S. P. Bulb-bearing Yellow Loosestrife. Northeast fourth of state; also in Lucas and Hardin Counties.
1374. Lysimachia nummularia L. Moneywort. General. Naturalized from Europe.
1375. Steironema ciliatum (L.) Raf. Fringed Yellow Loosestrife. General.
1376. Steironema lanceolatum (Walt.) Gr. Lanceleaf Yellow Loosestrife. Southwestern half of state.
1377. Steironema quadriflorum (Sims.) Hitch. Linear-leaf Yellow Loosestrife. Rather general.
1378. Naumburgia thyrsiflora (L.) Duby. Tufted Yellow Loosestrife. Rather general.
1379. Trientalis americana (Pers.) Pursh. Starflower. Cuyahoga, Portage, Summit, Champaign.
1380. Anagallis arvensis L. Scarlet Pimpernel. Lake, Fairfield, Logan, Montgomery, Greene, Scioto, Gallia. From Europe.
1381. Hottonia inflata Ell. Featherfoil. Ashtabula County.
1382. Samolus floribundus H. B. K. Water Pimpernel. Rather general.

1383. Dodecatheon meadia L. Shooting-star. Erie, Clark, Darke, Hocking, Hamilton, Clermont.

Plumbaginaceae. Leadwort Family.

1384. Ceratostigma plumbaginoides Bunge. Ceratostigma. A waif in Lake County.

Order, *Ericales*.

Pyrolaceae. Wintergreen Family.

1385. Pyrola americana Sw. Roundleaf Wintergreen. Defiance, Lucas, Lake, Wayne, Stark, Trumbull, Fairfield, Hocking.

1386. Pyrola elliptica Nutt. Shinleaf Wintergreen. Rather general, except in the southern part of Ohio.

1387. Pyrola secunda L. One-sided Wintergreen. Cuyahoga, Geauga, Summit, Portage.

1388. Chimaphila unbellata (L.) Nutt. Pipsissewa, Lucas, Cuyahoga, Portage.

1389 Chimaphila maculata (L.) Pursh. Spotted Pipsissewa. Rather general in eastern half of state.

Monotropaceae. Indian-pipe Family.

1390. Monotropa uniflora L. Indian-pipe. General.

1391. Hypopitys americana (DC.) Small. Smooth Pinesap. Fairfield, Hocking.

1392. Hypopitys lanuginosa (Mx) Nutt. Hairy Pinesap. Lake, Cuyahoga, Columbiana, Wayne.

Ericaceae. Heath Family.

1393. Ledum groenlandicum Oedr. Labrador Tea. Portage County.

1394. Azalea nudiflora L. Pink Azalea. Portage, Geauga, Lawrence.

1395. Azalea lutea L. Flame Azalea. Fairfield County.

1396. Azalea viscosa L. Swamp Azalea. Ashtabula County.

1397. Rhododendron maximum L. Great Rhododendron. Fairfield, Hocking.

1398. Kalmia latifolia L. Mountain Kalmia. Columbiana, Jefferson, Licking, Fairfield, Hocking, Jackson, Lawrence, Scioto.

1399 Chamaedaphne calyculata (L.) Moench. Leather-leaf. Geauga, Wayne, Stark, Williams, Defiance.

1400. Andromeda polifolia L. Wild Rosemary. Geauga, Wayne, Stark.

1401. Oxydendrum arboreum (L.) DC. Sorrel-tree. Adams, Fairfield, Hocking, Vinton, Jackson, Morgan, Meigs, Lawrence.

1402. Epigaea repens L. Trailing Arbutus. Cuyahoga, Geauga, Medina, Columbiana, Knox, Licking, Fairfield, Hocking, Jackson, Gallia, Lawrence.

1403. Gaultheria procumbens L. Creeping Wintergreen. Lucas, Cuyahoga, Wayne, Stark, Columbiana, Fairfield, Hocking, Jackson, Lawrence.

1404. Uva-ursi uva-ursi (L.) Britt. Bearberry. Erie County.

Vacciniaceae. Huckleberry Family.

1405. Polycodium stamineum (L.) Greene. Deerberry. Rather general in the eastern half of the state.

1406. Vaccinium corymbosum L. Tall Blueberry. Eastern half of state; also in Williams County.

1407. Vaccinium canadense Kalm. Canada Blueberry Lucas, Stark.

1408. Vaccinium angustifolium Ait. Dwarf Blueberry. Rather general in the eastern two-thirds of the state.

1409. Vaccinium vacillans Kalm. Low Blueberry. Eastern half of state; also in Fulton, Lucas and Ottawa Counties.

1410. Vaccinium atrococcum (Gr.) Heller. Dark Blueberry. Williams County.

1411. Chiogenes hispidula (L.) T. & G. Creeping Snowberry, Summit, Stark.

1412. Oxycoccus macrocarpus (Ait.) Pursh. Large Cranberry. Williams, Defiance, Geauga, Ashland, Wayne, Richland, Licking.

1413. Gaylussacia frondosa (L.) T. & G. Blue Huckleberry. No specimen.

1414. Gaylussacia baccata (Wang.) Koch. Black Huckleberry. General.

Order, *Ebenales.*

Ebenaceae. Ebony Family.

1415. Diospyros virginiana L. Persimmon. Southern half of the state; also in Lucas County.

Subclass, TUBIFLORAE.

Order, *Polemoniales.*

Polemoniaceae. Phlox Family.

1416. Phlox paniculata L. Garden Phlox. General.

1417. Phlox maculata L. Spotted Phlox. Rather general.

1418. Phlox ovata L. Mountain Phlox. Fulton County.

1419. Phlox glaberrima L. Smooth Phlox. No specimens.

1420. Phlox pilosa L. Downy Phlox. Northern part of the state, as far south as Franklin County.

1421. Phlox divaricata L. Wild Blue Phlox. General and abundant.

1422. Phlox stolonifera Sims. Creeping Phlox. Hocking County.

1423. Pholox subulata L. Ground Phlox. General.

1424. Gilia rubra (L.) Heller. Standing-cypress. Escaped in Erie and Lake Counties.

1425. Polemonium reptans L. Greek Valerian. General and abundant.

Convolvulaceae. Morning-glory Family.

1426. Ipomoea purpurea (L.) Lam. Common Morning-glory. General. From tropical America.

1427. Ipomoea hederacea Jacq. Ivyleaf Morning-glory. Southwestern Ohio, from Gallia to Auglaize County; also in Lake County. From tropical America.

1428. Ipomoea lacunosa L. Small-flowered White Morning-glory. Hamilton, Clermont, Gallia.

1429. Ipomoea pandurata (L.) Meyer. Wild Potato-vine. General.

1430. Quamoclit coccinea (L.) Moench. Small Red Morning-glory. No specimens. From tropical America.

1431. Convolvulus arvensis L. Small Bindweed. General. Naturalized from Europe.

1342. Convolvulus spithamaeus L. Upright Bindweed. Gallia, Clermont, Auglaize, Lucas, Portage.

1433. Convolvulus sepium L. Hedge Bindweed. General.

1434. Convolvulus japonicus Thunb. Japanese Bindweed. Fayette, Auglaize, Huron, Erie, Medina. Escaped.

Cuscutaceae. Dodder Family.

1435. Cuscuta epilinum Weihe. Flax Dodder. Wayne County. From Europe.

1436. Cuscuta epithymum Murr. Clover Dodder. Mercer County. From Europe.

1437. Cuscuta arvense Beyrich. Field Dodder. Vinton, Tuscarawas, Wayne, Erie.

1438. Cuscuta polygonorum Engel. Smartweed Dodder. Rather general.

1439. Cuscuta indecora Choicy. Pretty Dodder. Montgomery County.

1440. Cuscuta coryli Engel. Hazel Dodder. Rather general.

1441. Cuscuta cephalanthi Engel. Buttonbush Dodder. Ottawa. Franklin.

1442. Cuscuta gronovii Willd. Gronovius' Dodder. General.

1443. Cuscuta compacta Juss. Compact Dodder. No specimens.

1444. Cuscuta paradoxa Raf. Glomerate Dodder. No specimens. From the west.

Hydrophyllaceae. Water-leaf Family.

1445. Hydrophyllum virginianum L. Virginia Water-leaf. General.

1446. Hydrophyllum macrophyllum Nutt. Large Water-leaf. Western part of Ohio, as far east as Gallia, Fairfield, Licking, and Wyandot Counties.

1447. Hydrophyllum appendiculatum Mx. Appendaged Water-leaf. General.

1448. Hydrophyllum canadense L. Broadleaf Water-leaf. Cuyahoga, Wayne, Morrow, Hardin, Greene, Belmont.

1449. Phacelia dubia (L.) Small. Small-flowered Phacelia. Fairfield County.

1450. Phacelia bipinnatifida Mx. Loose-flowered Phacelia. Hamilton County.

1451. Phacelia purshii Buckl. Pursh's Phacelia. General.

Order, *Gentianales.*

Loganiaceae. Logania Family.

1452. Spigelia marilandica L. Indian-pink. Lake County.

Oleaceae. Olive Family.

1453. Syringa vugaris L. Common Lilac. Lake, Jefferson. Escaped.

1454. Ligustrum vulgare L. Privet. Rather general. From Europe.

1455. Chionanthus virginica L. Fringetree. Meigs, Gallia, Pike.

1456. Fraxinus nigra Marsh. Black Ash. General in the northern part of the state, south to Preble, Green, Franklin and Harrison Counties.

1457. Fraxinus quadrangulata Mx. Blue Ash. Erie, Ottawa, Hancock, Auglaize, Franklin, Licking, Montgomery, Highland, Ross, Brown, Adams.

1458. Fraxinus pennsylvanica Marsh. Red Ash. General.

1459. Fraxinus lanceolata Borck. Green Ash. General.

1460. Fraximus biltmoreana Beadle. Biltmore Ash. Erie, Hardin, Franklin, Montgomery, Morgan, Hamilton, Brown, Lawrence, Meigs.

1461. Fraxinus americana L. White Ash. General and abundant.

Gentianaceae. Gentian Family

1462. Centaurium centaurium (L.) Wight. European Centaury. Lake County. From Europe.

1463. Sabbatia angularis (L.) Pursh. Square-stemmed Sabbatia. General.

1464. Gentiana quinquefolia L. Stiff Gentian. Montgomery, Ross., Franklin, Wayne, Summit, Clark, Fulton, Cuyahoga.

1465. Gentiana crinita Froel. Fringed Gentian. Cuyahoga, Erie, Fulton, Auglaize, Champaign, Clark, Madison, Franklin.

1466. Gentiana puberula Mx. Downy Gentian. (Dasystephana). Erie County.

1467. Gentiana saponaria L. Soapwort Gentian. (Dasystephana). Cuyahoga, Lucas.

1468. Gentiana andrewsii Griseb. Closed Gentian. (Dasystephana). General, but no specimens southeast of Hamilton, Fairfield, and Columbiana Counties.

1469. Gentiana flavida Gr. Yellowish Gentian. (Dasystephana). Lucas County.

1470. Gentiana villosa L. Striped Gentian. (Dasystephana). Gallia County.

1471. Frasera carolinensis Walt. American Columbo. From Madison County southward and westward; also in Summit, Geauga and Portage Counties.

1472. Obolaria virginica L. Pennywort. Lake, Cuyahoga, Fairfield, Gallia, Lawrence, Clermont, Hamilton.

1473. Bartonia virginica (L.) B. S. P. Yellow Bartonia. Licking, Lake.

Menyanthaceae. Buckbean Family.

1474. Menyanthes trifoliata L. Buckbean. Ashtabula, Wayne, Licking.

Apocynaceae. Dogbane Family.

1475. Vinca minor L. Periwinkle. Rather general. Native of Europe.
1476. Apocynum androsaemifolium L. Spreading Dogbane. General.
1477. Apocynum cannabinum L. Indian Hemp. General and abundant.
1478. Apocynum sibiricum Jacq. Clasping-leaf Dogbane. Erie, Ashtabula.
1479. Apocynum pubescens R. Br. Velvet Dogbane. Auglaize, Franklin, Harrison, Adams.

Asclepiadaceae. Milkweed Family.

1480. Acerates viridiflora (Raf.) Eat. Green Milkweed. Rather general.
1481. Acerates floridana (Lam.) Hitch. Florida Milkweed. Erie, Jackson, Gallia.
1482. Asclepias tuberosa L. Pleurisy-root. General.
1483. Asclepias decumbens L. Decumbent Pleurisy-root. No specimens.
1484. Asclepias purpurascens L. Purple Milkweed. Madison, Auglaize, Lucas, Summit, Portage, Stark, Carroll.
1485. Asclepias incarnata L. Swamp Milkweed. General and abundant.
1486. Asclepias pulchra Ehrh. Hairy Milkweed. Lorain County.
1487. Asclepias sullivantii Engel. Sullivant's Milkweed. Erie, Fairfield.
1488. Asclepias amplexicaulis Sm. Bluntleaf Milkweed. Erie, Fairfield.
1489. Asclepias exaltata (L.) Muhl. Tall Milkweed. General.
1490. Asclepias variegata L. White Milkweed. Summit, Hocking.
1491. Asclepias quadrifolia L. Fourleaf Milkweed. General.
1492. Asclepias syriaca L. Common Milkweed. General and abundant.
1493. Asclepias verticillata L. Whorled Milkweed. Lucas, Ottawa, Erie, Cuyahoga, Clark, Greene, Fairfield, Athens.

1494. Gonolobus laevis Mx. Sandvine. Southern Ohio, as far north as Montgomery, Ross, and Washington Counties.

1495. Cynanchum nigrum (L.) Pers. Black Swallow-wort. Cuyahoga, Lake. From Europe.

1496. Vincetoxicum obliquum (Jacq.) Britt. Large-flowered Vincetoxicum. Lawrence, Hamilton, Greene, Franklin.

Order, *Scrophulariales.*

Solanaceae. Potato Family.

1497. Petunia violacea Lindl. Common Petunia. Monroe, Franklin. From South America.

1498. Nicotiana tabacum L. Common Tobacco. Adams, Huron. Escaped from cultivation.

1499. Datura metel L. Entire-leaf Jimson-weed. Lake County. From tropical America.

1500. Datura stramonium L. Common Jimson-weed. General. Naturalized. From the tropics.

1501. Lycium halmifolium Mill. Matrimony-vine. Rather general. From Europe.

1502. Physalodes physalodes (L.) Britt. Apple-of-Peru. Gallia, Hamilton, Clinton, Montgomery, Clark, Champaign, Franklin, Licking. From Peru.

1503. Physalis lanceolata Mx. Prairie Ground-cherry. General.

1504. Physalis ixocarpa Brot. Mexican Ground-cherry Franklin County. Spontaneous after cultivation.

1505. Physalis virginiana Mill. Virginia Ground-cherry. Cuyahoga County.

1506. Physalis alkekengi L. Chinese-lantern (Ground-cherry). Persistent after cultivation. Franklin, Lake.

1507. Physalis heterophylla Nees. Clammy Ground-cherry. General and abundant.

1508. Physalis pubescens L. Low Hairy Ground-cherry. Morgan, Shelby.

1509. Physalis pruinosa L. Tall Hairy Ground-cherry. Franklin County. Escaped from cultivation.

1510. Solanum elaeagnifolium Cav. Silver-leaf Nightshade. A waif in Lucas County.

1511. Solanum carolinense L. Horse-nettle. General.

1512. Solanum tuberosum L. Potato. Monroe, Hocking, Franklin, Tuscarawas, Erie, Ottawa. Persistent after cultivation.

1513. Solanum dulcamara L. Bittersweet. General as far south as Clark, Licking, and Jefferson Counties; also in Meigs County. Naturalized from Europe.

1514. Solanum nigrum L. Black Nightshade. General and abundant.

1515. Solanum rostratum Dun. Buffalo-bur. Franklin, Marion, Ottawa, Cuyahoga, Summit, Lake. From the West.

1516. Lycopersicon lycopersicon (L.) Karst. Tomato. Rather general as an escape.

Scrophulariaceae. Figwort Family.

1517. Verbascum blattaria L. Moth Mullen. General and abundant. Naturalized from Europe.

1518. Verbascum thapsus L. Common Mullen. General and abundant. Naturalized from Europe.

1519. Scrophularia leporella Bickn. Hare Figwort. Cuyahoga County.

1521. Chelone glabra L. Smooth Turtle-head. General.

1522. Pentstemon hirsutus (L.) Willd. Hairy Beard-tongue. General.

1523. Pentstemon pentstemon (L.) Britt. Smooth Beard-tongue. Rather general.

1524. Pentstemon digitalis (Sweet) Nutt. Foxglove Beard-tongue. Rather general.

1525. Pentstemon cobaea Nutt. Cobaea Beard-tongue. Lake County.

1526. Collinsia verna Nutt. Blue-eyed-Mary. General.

1527. Mimulus ringens L. Square-stemmed Monkey-flower. General.

1528. Mimulus alatus Soland. Sharp-winged Monkey-flower. Rather general.

1529. Conobea multifida (Mx.) Benth. Conobea. Hamilton, Greene, Madison, Ottawa.

1530. Gratiola virginiana L. Clammy Hedge-hyssop. General.

1531. Gratiola sphaerocarpa Ell. Round-fruited Hedge-hyssop. Erie County.

1532. Ilysanthes dubia (L.) Barnh. Long-stalked False Pimpernel. Rather general.

1533. Ilysanthes attenuata (Muhl.) Small. Short-stalked False Pimpernel. Cuyahoga, Lorain, Huron, Stark, Scioto.

1534. Synthysis bullii (Eat.) Heller. Bull's Synthyris. Montgomery County.

1535. Veronica anagallis-aquatica L. Water Speedwell. Butler, Champaign, Auglaize, Lucas, Erie.

1536. Veronica americana Schwein. American Speedwell. Rather general.

1537. Veronica scutellata L. Skullcap Speedwell. Franklin, Licking, Crawford, Perry, Lucas, Ottawa, Erie, Cuyahoga.

1538. Veronica officinalis L. Common Speedwell. General and abundant.

1539. Veronica chamaedris L. Bird's-eye Speedwell. Lake County. From Europe.

1540. Veronica teucrium L. Germander Speedwell. Medina County. From Europe.

1541. Veronica serpyllifolia L. Thyme-leaf Speedwell. General.

1542. Veronica peregrina L. Purslane Speedwell. General.

1543. Veronica arvensis L. Field Speedwell. General. From Europe.

1544. Veronica agrestis L. Garden Speedwell. Montgomery, Franklin. From Europe.

1545. Veronica tournefortii Gmel. Tournefort's Speedwell. Madison, Franklin, Jefferson, Lorain, Cuyahoga, Lake. From Europe.

1546. Veronica hederaefolia L. Ivyleaf Speedwell. (Erie County—Moseley Herbarium.) From Europe.

1547. Leptandra virginica (L.) Nutt. Culver's-root. General.

1548. Digitalis purpurea L. Purple Foxglove. Cuyahoga, Lake. From Europe.

1549. Digitalis lutea L. Yellow Foxglove. A waif in Cuyahoga County.

1550. Buchnera americana L. Blue-hearts. Fulton County.

1551. Afzelia macrophylla (Nutt.) Ktz. Mullen Foxglove. General in western Ohio, as far east as Huron, Noble, and Vinton Counties.

1552. Dasystoma pedicularia (L.) Benth. Fernleaf False Foxglove. Fulton County.

1553. Dasystoma flava (L.) Wood. Downy False Foxglove. Eastern Ohio, as far west as Cuyahoga, Fairfield, and Adams Counties.

1554. Dasystoma laevigata Raf. Entire-leaf False Foxglove. Jackson, Vinton, Hocking, Fairfield.

1555. Dasystoma virginica (L.) Britt. Smooth False Foxglove. Fulton, Wood, Fairfield, Adams.

1556. Agalinis purpurea (L.) Britt. Large Purple Gerardia. Montgomery, Fairfield, Franklin, Fulton, Erie, Wayne.

1557. Agalinis paupercula (Gr.) Britt. Small Purple Gerardia. Stark, Ottawa, Logan, Champaign, Gallia.

1558. Agalinis tenuifolia (Vahl.) Raf. Slender Gerardia. General.

1559. Agalinis skinneriana (Wood) Britt. Skinner's Gerardia. Greene County.

1560. Otophylla auriculata (Mx.) Small. Auricled Gerardia. Ottawa County.

1561. Castilleja coccinea (L.) Spreng. Scarlet Painted-cup. Franklin, Knox.

1562. Pedicularis lanceolata Mx. Lanceleaf Lousewort. Rather general, but no specimens from south of Montgomery and Hocking Counties.

1563. Pedicularis canadensis L. Wood Lousewort. General.

1564. Melampyrum lineare Lam. Narrow-leaf Cow-wheat. Lorain, Cuyahoga, Lake, Ashtabula, Geauga, Portage, Hocking.

1565. Antirrhinum majus L. Great Snapdragon. Madison County. From Europe.

1566. Linaria linaria (L.) Karst. Yellow Toadflax. General, but no specimens from the northwestern counties. Naturalized from Europe.

1657. Linaria canadensis (L.) Dum. Blue Toadflax. No specimens.

1568. Chaenorrhinum minus (L.) Lange. Lesser Toadflax. Portage County. From Europe.

1569. Kickxia spuria (L.) Dum. Roundleaf Toadflax. Lake County. From Europe.

1570. Kickxia elatine (L.) Dum. Sharp-pointed Toadflax. Lake County. From Europe.

1571. Cymbalaria cymbalaria (L.) Wettst. Kenilworth Ivy. Montgomery, Crawford. From Europe.

Orobanchaceae. Broom-rape Family.

1572. Thalesia uniflora (L.) Britt. Naked Broom-rape. Rather general, but no specimens from the northwestern counties.

1573. Orobanche ludoviciana Nutt. Louisiana Broom-rape. Hamilton County.

1574. Conopholis americana (L. f.) Wallr. Squaw-root. General.

1575. Leptamnium virginianum (L.) Raf. Beech-drops. General.

Bignoniaceae. Bignonia Family.

1576. Bignonia radicans L. Trumpet-creeper. General.
1577. Anisostichus capreolata (L.) Bur. Cross-vine. Lawrence, Adams.
1578. Catalpa catalpa (L.) Karst. Common Catalpa. Montgomery, Champaign, Franklin. From the South.
1579. Catalpa speciosa Ward. Hardy Catalpa. Ashtabula, Franklin, Hocking, Madison, Preble. From the Southwest.

Martyniaceae. Unicorn-plant Family.

1580. Martynia louisiana Mill. Unicorn-plant. Lorain, Richland, Franklin, Ross. Escaped from cultivation.

Lentibulariaceae. Bladderwort Family.

1581. Utricularia macrorhiza Le Conte. Greater Bladderwort. Northern part of state; specimens from as far south as Franklin and Licking Counties.
1582. Utricularia gibba L. Humped Bladderwort. Stark, Franklin, Fairfield, Erie, Defiance.
1583. Utricularia intermedia Hayne. Flatleaf Bladderwort. Lake, Wayne.
1584. Utricularia minor L. Lesser Bladderwort. Licking County.
1585. Stomoisia cornuta (Mx.) Raf. Horned Bladderwort. Summit County.

Acanthaceae. Acanthus Family.

1586. Ruellia strepens L. Smooth Ruellia. Western half of the state; also in Monroe County.
1587. Ruellia ciliosa Pursh. Hairy Ruellia. From Union County southward and westward; also in Cuyahoga County.
1588. Dianthera americana L. Water-willow. General.

Order, *Lamiales.*

Boraginaceae. Borage Family.

1589. Heliotropium indicum L. Indian Heliotrope. No specimens. From India.
1590. Cynoglossum officinale L. Hound's-tongue. General and abundant. Naturalized from Europe.
1591. Cynoglossum virginianum L. Wild Comfrey. Southeastern Ohio to Cuyahoga, Wyandot, and Warren Counties.

1592. Lappula lappula (L.) Karst. European Stickseed. Rather general. From Europe.

1593. Lappula virginiana (L.) Greene. Virginia Stickseed. General.

1594. Mertensia virginica (L.) DC. Virginia Cowslip. Rather general.

1595. Asperugo procumbens L. German Madwort. Lake County. From Europe.

1596. Myosotis laxa Lehm. Smaller Forget-me-not. Tuscarawas, Stark, Summit, Lake, Lucas.

1597. Myosotis arvensis (L.) Hill. Field Forget-me-not. Gallia, Franklin, Lake.

1598. Myosotis virginica (L.) B. S. P. Virginia Forget-me-not. Fairfield, Franklin, Lorain, Erie, Lucas.

1599. Lithospermum latifolium Mx. American Gromwell. Lawrence, Warren, Lucas, Auglaize.

1600. Lithospermum officinale L. Common Gromwell. No specimens. From Europe.

1601. Lithospermum arvense L. Corn Gromwell. General and abundant. Naturalized from Europe.

1602. Lithospermum carolinense (Walt.) MacM. Hairy Puccoon. Erie, Lucas.

1603. Lithospermum canescens (Mx.) Lehm. Hoary Puccoon. Rather general; no specimens from the eastern counties.

1604. Onosmodium hispidissimum Mack. Shaggy False Gromwell. Lorain, Ottawa, Lucas, Greene, Clark, Montgomery, Adams.

1605. Symphytum officinale L. Common Comfrey. Northern part of Ohio; as far south as Belmont and Champaign Counties. From Europe.

1606. Lycopsis arvensis L. Small Bugloss. Stark County. From Europe.

1607. Echium vulgare L. Blueweed. Montgomery, Clinton, Richland, Cuyahoga, Columbiana, Noble, Belmont. From Europe.

Verbenaceae. Vervain Family.

1608. Verbena urticifolia L. White Vervain. General and abundant.

1609. Verbena hastata L. Blue Vervain. General and abundant.

1610. Verbena angustifolia Mx. Narrowleaf Vervain. Lake, Cuyahoga, Erie, Ottawa, Auglaize, Madison, Montgomery, Clermont, Adams, Meigs.

1611. Verbena stricta Vent. Hoary Vervain. Hamilton, Clermont, Highland, Warren, Preble, Montgomery, Clark, Franklin, Licking, Cuyahoga, Lake.

1612. Verbena bracteosa Mx. Bracted Vervain. Pike, Hamilton, Montgomery, Auglaize, Wyandot, Cuyahoga, Ashtabula.

1613. Verbena canadensis (L.) Britt. Large-flowered Verbena. Ross, Franklin, Auglaize.

1614. Lippia lanceolate Mx. Frog-fruit. General; no specimens from the eastern counties.

Lamiaceae. Mint Family.

1615. Isanthus brachiatus (L.) B. S. P. False Pennyroyal. Erie, Ottawa, Clark, Warren, Franklin, Muskingum, Hocking, Morgan, Gallia.

1616. Trichostema dichotomum L. Blue-curls. Hamilton, Fairfield, Jackson, Monroe, Highland.

1617. Ajuga reptans L. Bungle-weed. No specimens. From Europe.

1618. Teucrium canadense L. American Germander. General and abundant.

1619. Teucrium occidentale Gr. Hairy Germander. Lake, Wayne, Erie, Ottawa, Auglaize, Clark, Greene, Perry, Monroe.

1620. Teucrium scorodonia L. Wood Germander. Lake County. From Europe.

1621. Teucrium botrys L. Cutleaf Germander. No specimens. From Europe.

1622. Scutellaria lateriflora L. Mad-dog Skullcap. General.

1623. Scutellaria serrata Andr. Showy Skullcap. Gallia County.

1624. Scutellaria incana Muhl. Downy Skullcap. Eastern and southern Ohio; from Cuyahoga and Wayne Counties eastward and from Perry and Miami Counties southward.

1625. Scutellaria cordifolia Muhl. Heartleaf Skullcap. Rather general.

1626. Scutellaria pilosa Mx. Hairy Skullcap. No specimens.

1627. Scutellaria integrifolia L. Hyssop Skullcap. Jackson County.

1628. Scutellaria parvula Mx. Small Skullcap. Ottawa, Madison, Clark, Hamilton, Gallia, Franklin, Greene, Montgomery, Scioto.

1629. Scutellaria saxatilis Ridd. Rock Skullcap. No specimens.

1630. Scutellaria galericulata L. Marsh Skullcap. Rather general; no specimens south of Clark and Perry Counties.

1631. Scutellaria nervosa Pursh. Veined Skullcap. Rather general.

1632. Marrubium vulgare L. Common Hoarhound. General. Naturalized from Europe.

1633. Hedeoma pulegioides (L.) Pers. American Pennyroyal. General and abundant.

1634. Hedeoma hispida Pursh. Rough Pennyroyal. (Moseley Herbarium—Lorain County.)

1635. Melissa officinalis L. Lemon Balm. Rather general; no specimens from the northwestern counties. From Europe.

1636. Satureia hortensis L. Summer Savory. Ottawa, Lake. From Europe.

1637. Clinopodium vulgare L. Field Basil. Rather general; no specimens from southeastern and northwestern counties.

1638. Clinopodium glabrum (Nutt.) Ktz. Low Calamint. Ottawa, Erie, Union, Greene.

1639. Koellia virginiana (L.) MacM. Virginia Mountain-mint. Rather general.

1640. Koellia flexuosa (Walt.) MacM. Narrowleaf Mountain-mint. Rather general.

1641. Koellia pilosa (Nutt.) Britt. Hairy Mountain-mint. Cuyahoga, Stark, Hocking, Clark, Shelby.

1642. Koellia incana (L.) Ktz. Hoary Mountain-mint. Stark, Fairfield, Hocking, Jackson, Gallia, Scioto, Adams.

1643. Koellia mutica (L.) Britt. Short-toothed Mountain-mint. Licking, Cuyahoga.

1644. Origanum vulgare L. Wild Majoram. Hocking County. From Europe.

1645. Thymus serpyllum L. Creeping Thyme. Coshocton, Gallia. From Europe.

1646. Cunila origanoides (L.) Britt. American Dittany. Southeastern Ohio to Tuscarawas, Fairfield, Ross, and Adams Counties.

1647. Lycopus virginicus L. Virginia Water-hoarhound. Rather general.

1648. Lycopus uniflorus Mx. Northern Water-hoarhound. Belmont County.

1649. Lycopus rubellus Moench. Stalked Water-hoarhound. Geauga, Wayne, Cuyahoga, Erie, Huron, Paulding, Auglaize, Fairfield, Clinton, Montgomery.

1650. Lycopus americanus Muhl. Cutleaf Water-hoarhound. General

1651. Mentha spicata L. Spearmint. General. Naturalized from Europe.
1652. Mentha piperita L. Peppermint. General. Naturalized from Europe.
1653. Mentha citrata Ehrh. Bergamot Mint. Lake, Franklin. From Europe.
1654. Mentha longifolia (L.) Huds. European Mint. Lake County. From Europe.
1655. Mentha rotundifolia (L.) Huds. Roundleaf Mint. Franklin County—a waif. From Europe.
1656. Mentha alopecuroides Hull. Woolly Mint. Franklin County. From Europe.
1657. Mentha arvensis L. Field Mint. Lake County. From Europe.
1658. Mentha cardiaca Gerarde. Small-leaf Mint. Montgomery County. From Europe.
1659. Mentha canadensis L. American Wild Mint. General.
1660. Collinsonia canadensis L. Stone-root. General.
1661. Perilla frutescens (L.) Britt. Perilla. Warren County. Native of India.
1662. Agastache nepetoides (L.) Ktz. Catnip Giant-hyssop. General.
1663. Agastache scrophulariaefolia (Willd.) Ktz. Figwort Giant-hyssop. Medina, Stark, Auglaize, Champaign, Madison, Hocking.
1664. Nepeta cataria L. Catnip. General and abundant. Naturalized from Europe.
1665. Glecoma hederacea L. Ground Ivy. General and abundant. Naturalized from Europe.
1666. Prunella vulgaris L. Common Self-heal. General and abundant. Native of Europe.
1667. Dracocephalum virginianum L. Virginia Dragon-head. Rather general.
1668. Synandra hispidula (Mx.) Britt. Synandra. Belmont, Wyandot, Franklin, Miami, Hamilton, Clermont, Lawrence.
1669. Galeopsis tetrahit L. Hemp-nettle. Lake County. From Europe.
1670. Leonurus cardiaca L. Common Motherwort. General and abundant. Naturalized from Europe.
1671. Lamium amplexicaule L. Common Henbit. Rather general. From Europe.

1672. Lamium purpureum L. Red Henbit. Warren, Erie, Lorain. From Europe.
1673. Lamium maculatum L. Spotted Henbit. Washington, Miami, Knox, Marion, Auglaize, Lorain. From Europe.
1674. Lamium album L. White Henbit. Miami, Lorain. From Europe.
1675. Stachys palustris L. Marsh Hedge-nettle. Rather general.
1676. Stachys tenuifolia Willd. Smooth Hedge-nettle. Rather general.
1677. Stachys asper Mx. Rough Hedge-nettle. Rather general.
1678. Stachys cordata Ridd. Cordate Hedge-nettle. Southern Ohio, as far north as Noble, Franklin, and Auglaize Counties.
1679. Blephilia ciliata (L.) Raf. Downy Blephilia. Northern part of the state; as far south as Harrison, Franklin, and Montgomery Counties.
1680. Blephilia hirsuta (Prush) Torr. Hairy Blephilia. General.
1681. Monarda punctata L. Horsemint. No specimens.
1682. Monarda didyma L. American Beebalm. Trumbull, Portage, Cuyahoga, Medina, Madison.
1683. Monarda clinopodia L. Basil Balm. General.
1684. Monarda fistulosa L. Wild Bergamot. General and abundant.
1685. Monarda mollis L. Canescent Wild Bergamot. Rather general.
1686. Salvia lyrata L. Lyreleaf Sage. Pike, Lawrence, Gallia, Meigs.
1687. Salvia lanceifolia Poir. Lanceleaf Sage. Franklin County. From the West.
1688. Salvia verbenaca L. Wild Sage. No specimens. From Europe.
1689. Salvia officinalis L. Common Sage. Stark County. Escaped.

Phrymaceae. Lopseed Family.

1690. Phryma leptostachya L. Lopseed. General.

Order, *Plantaginales.*

Plantaginaceae. Plaintain Family.

1691. Plantago cordata Lam. Heartleaf Plantain. Lucas, Auglaize, Madison, Franklin.
1692. Plantago rugellii Dec. Rugel's Plantain. General and abundant.
1693. Plantago major L. Common Plantain. Rather general.
1694. Plantago lanceolata L. Ribgrass Plantain. General and abundant. Naturalized from Europe.
1695. Plantago aristata Mx. Large-bracted Plantain. Rather general.

1696. Plantago virginica L. Dwarf Plantain. Gallia, Jackson, Pike, Ross, Stark, Cuyahoga, Lake.

1697. Plantago arenaria W. & K. Sand Plantain. Montgomery County. From Europe.

Subclass, INFERAE.

Order, *Umbellales.*

Araliaceae. Ginseng Family.

1698. Aralia spinosa L. Angelica-tree. Clermont, Hocking.

1699. Aralia racemosa L. American Spikenard. General.

1700. Aralia nudicaulis L. Wild Sarsaparilla. Northern part of state, as far south as Licking County.

1701. Aralia hispida Vent. Bristly Sarsaparilla. Lake, Cuyahoga.

1702. Panax quinquefolium L. Common Ginseng. General.

1703. Panax trifolium L. Dwarf Ginseng. Columbiana, Cuyahoga, Lorain, Medina, Seneca, Richland.

Ammiaceae. Carrot Family.

1704. Eryngium aquaticum L. Rattlesnake-master. Wyandot, Erie.

1705. Sanicula marylandica L. Black Snakeroot. Rather general.

1706. Sanicula gregaria Bickn. Clustered Snakeroot. Rather general

1707. Sanicula canadensis L. Short-styled Snakeroot. General and abundant.

1708. Sanicula trifoliata Bickn. Large-fruited Snakeroot. Southern and eastern part of state to Geauga, Morrow, and Preble Counties.

1709. Deringa canadensis (L.) Ktz. Honewort. General.

1710. Chaerophyllum procumbens (L.) Crantz. Spreading Chervil. General.

1711. Washingtonia claytoni (Mx.) Britt. Woolly Sweet-cicely. General.

1712. Washingtonia longistylis (Torr.) Britt. Long-styled Sweet-cicely. General.

1713. Scandix pecten-veneris L. Venus'-comb. Lake County. From Europe.

1714. Pastinaca sativa L. Wild Parsnip. General and abundant. Naturalized from Europe.

1715. Heracleum lanatum Mx. Cow-parsnip. Rather general.

1716. Conioselium chinense (L.) B. S. P. Hemlock-parsley. Lake, Summit.

1717. Angelica atropurpurea L. Purple-stemmed Angelica. Rather general.

1718. Angelica villosa (Walt.) B. S. P. Pubescent Angelica. Eastern half of state, west to Adams, Fairfield and Richland Counties.

1719. Oxypolis rigidus (L.) Raf. Cowbane. Hamilton, Clark, Franklin, Erie, Fulton, Champaign, Huron, Montgomery, Madison.

1720. Bupleurum rotundifolium L. Hare's-ear. Warren County. From Europe.

1721. Thaspium trifoliatum (L.) Britt. Purple Meadow-parsnip. General.

1722. Thaspium barbinode (Mx.) Nutt. Hairy-jointed Meadow-parsnip. General.

1723. Taenidia integerrima (L.) Drude. Yellow Pimpernel. General.

1724. Zizia aurea (L.) Koch. Early Meadow-parsnip. Rather general.

1725. Zizia cordata (Walt.) DC. Heartleaf Meadow-parsnip. Lorain, Richland, Wyandot, Madison, Franklin, Warren, Gallia, Washington.

1726. Apium petroselinum L. Parsley. Madison County. Escaped from cultivation.

1727. Foeniculum foeniculum (L.) Karst. Fennel. Hocking, Scioto. From Europe.

1728. Aethusa cynapium L. Fool's Parsley. Lake County. From Europe.

1729. Hydrocotyle umbellata L. Umbellate Marsh-pennywort. Portage, Stark.

1730. Hydrocotyle americana L. American Marsh-pennywort. Cuyahoga, Summit, Wayne, Stark.

1731. Erigenia bulbosa (Mx.) Nutt. Harbenger-of-spring. General.

1732. Conium maculatum L. Poison-hemlock. Montgomery, Knox, Lake. From Europe.

1733. Aegopodium podagraria L. Goutweed. Lake County. From Europe.

1734. Eulophus americanus Nutt. Eastern Eulophus. No specimens.

1735. Pimpinella saxifraga L. Pimpernel. No specimens. From Europe.

1736. Sium cicutaefolium Schrank. Water-parsnip. General, but no specimens from the southern counties.

1737. Cicuta maculata L. Spotted Water-hemlock. General.

1738. Cicuta bulbifera L. Bulb-bearing Water-hemlock. Northern part of state south to Perry and Clark Counties.

1739. Carum carui L. Caraway. Columbiana, Ashland, Lorain, Fulton. From Europe.

1740. Daucus carota L. Wild Carrot. General and abundant. Naturalized from Europe.

1741. Torilis anthriscus (L.) Gmel. Erect Hedge-parsley. Hamilton County. From Europe.

Cornaceae. Dogwood Family.

1742. Cornus alternifolia L. f. Blue Dogwood. General.

1743. Cornus femina Mill. Panicled Dogwood. General in the northern half of the state.

1744. Cornus stolonifera Mx. Red-osier Dogwood. General in the northern part of the state, south to Stark, Morrow, and Montgomery Counties.

1745. Cornus asperifolia Mx. Roughleaf Dogwood. General.

1746. Cornus amomum Mill. Silky Dogwood. General and abundant.

1747. Cornus rugosa Lam. Roundleaf Dogwood. Cuyahoga, Summit, Warren.

1748. Cynoxylon floridum (L.) Raf. Flowering Dogwood. General and abundant.

1749. Cynoxylon canadense (L.) Dwarf Dogwood. Stark, Licking.

1750. Nyssa sylvatica Marsh. Tupelo. General.

Order, *Rubiales.*

Rubiaceae. Madder Family.

1751. Houstonia coerulea L. Bluets. Southeastern two-thirds of the state as far northwest as Cuyahoga, Crawford, Clark, and Hamilton Counties.

1752. Houstonia purpurea L. Large Houstonia. Clermont, Butler, Highland, Warren.

1753. Houstonia ciliolata Torr. Fringed Houstonia. Lawrence, Licking, Franklin, Delaware, Defiance, Lucas, Ottawa, Cuyahoga, Lake.

1754. Houstonia longifolia Gaertn. Longleaf Houstonia. Rather general, but Ottawa the only northern county represented in the herbarium.

1755. Houstonia tenuifolia Nutt. Slenderleaf Houstonia. No specimens.

1756. Houstonia angustifolia Mx. Narrowleaf Houstonia. Ottawa County.

1757. Cephalanthus occidentalis L. Button-bush. General and abundant.

1758. Michella repens L. Partridge-berry. Rather general.

1759. Spermacoce glabra Mx. Smooth Buttonweed. No specimens.

1760. Diodia teres Walt. Rough Buttonweed. Lake County.

1761. Galium pilosum Ait. Hairy Bedstraw. Eastern Ohio; as far west as Lorain, Knox, Fairfield, and Adams Counties.

1762. Galium lanceolatum Torr. Lanceleaf Wild Licorice. Rather general.

1763. Galium circaezans Mx. Wild Licorice. General and abundant.

1764. Galium boreale L. Northern Bedstraw. Ottawa, Lorain.

1765. Galium triflorum Mx. Fragrant Bedstraw. General.

1766. Galium mollugo L. White Bedstraw. Lake, Fayette. From Europe.

1767. Galium tinctorium L. Stiff Marsh Bedstraw. Rather general.

1768. Galium trifidum L. Small Bedstraw. Northern Ohio, as far south as Shelby, Madison, Perry, and Harrison Counties.

1769. Galium claytoni Mx. Clayton's Bedstraw. Erie County.

1770. Galium concinnum T. & G. Shining Bedstraw. General and abundant.

1771. Galium asprellum Mx. Rough Bedstraw. Rather general; no specimens from the southern counties.

1772. Galium aparine L. Common Cleavers. General and abundant.

1773. Sherardia arvensis L. Blue Field-madder. Cuyahoga County. From Europe.

Caprifoliaceae. Honeysuckle Family.

1774. Sambucus canadensis L. Common Elderberry. General and abundant.

1775. Sambucus racemosa L. Red Elderberry. Rather general.

1776. Viburnum pubescens (Ait.) Pursh. Downy Arrow-wood. Lorain, Erie, Wyandot, Auglaize, Williams.

1777. Viburnum dentatum L. Toothed Arrow-wood. Ashtabula, Geauga, Lorain, Summit, Stark, Wayne, Ashland, Tuscarawas.

1778. Viburnum scabrellum (T. & G.) Chapm. Roughleaf Arrow-wood. Adams, Brown, Hocking, Madison.

1779. Viburnum cassinoides L. Withe-rod. Ashtabula, Geauga, Cuyahoga, Summit, Lorain, Hocking.

1780. Viburnum lentago L. Sheepberry. Rather general.

1781. Viburnum prunifolium L. Black Haw. General.

1782. Viburnum lantana L. Wayfaring-tree. Lake County. From Europe.

1783. Viburnum acerifolium L. Mapleleaf Arrow-wood. General.

1784. Viburnum opulus L. Cranberry-tree. Lake, Geauga, Champaign.

1785. Viburnum alnifolium Marsh. Hobblebush. Ashtabula, Lake.

1786. Triosteum angustifolium L. Yellow Horse-gentian. Cuyahoga, Warren, Clermont.

1787. Triosteum perfoliatum L. Common Horse-gentian. General.

1788. Symphoricarpos racemosus Mx. Snowberry. Rather general.

1789. Symphoricarpos symphoricarpos (L.) MacM. Coralberry. General.

1790. Lonicera canadensis Marsh. American Fly Honeysuckle. Lake, Summit, Cuyahoga, Lorain.

1791. Lonicera oblongifolia (Goldie) Hook. Swamp Fly Honeysuckle. Cuyahoga County.

1792. Lonicera tartarica L. Tartarian Honeysuckle. Ashtabula, Lake, Cuyahoga, Lorain, Licking, Franklin, Auglaize. Escaped from cultivation.

1793. Lonicera xylosteum L. European Fly Honeysuckle. Lake County. Native of Europe.

1794. Lonicera japonica Thunb. Japanese Honeysuckle. Adams, Brown, Auglaize. Escaped from cultivation.

1795. Lonicera sempervirens L. Trumpet Honeysuckle. Cuyahoga County.

1796. Lonicera caprifolium L. Italian Honeysuckle. No specimens. From Europe.

1797. Lonicera hirsuta Eaton. Hairy Honeysuckle. Ottawa, Lorain, Monroe.

1798. Lonicera glaucescens Rydb. Glauscent Honeysuckle. General.

1799. Lonicera sullivantii Gr. Sullivant's Honeysuckle. Stark, Muskingum, Franklin, Madison, Clark, Highland.

1800. Lonicera dioica L. Smoothleaf Honeysuckle. Champaign, Franklin.

1801. Linnaea americana Forbes. American Twin-flower. Stark County.

1802. Diervilla diervilla (L.) MacM. Bush-honeysuckle. Lucas, Lorain, Summit, Wayne, Stark, Franklin.

Valerianaceae. Valerian Family

1803. Valerianella locusta (L.) Bettke. European Corn-salad. Hamilton, Ross, Lorain, Cuyahoga. From Europe.

1804. Valerianella chenopodifolia (Pursh.) DC. Goosefoot Corn-salad. Rather general.

1805. Valerianella radiata (L.) Dufr. Beaked Corn-salad. General.

1806. Valerianella woodsiana (T. & G.) Walp. Wood's Corn-salad. Erie, Richland, Franklin, Clark.

1807. Valeriana pauciflora Mx. Large-flowered Valerian. Western half of Ohio, as far east as Ottawa, Franklin, and Lawrence Counties.

1808. Valeriana edulis Nutt. Edible Valerian. Champaign County.

1809. Valeriana officinalis L. Garden Valerian. Ashtabula, Lake. From Europe.

Order, *Campanulales.*

Campanulaceae. Bellflower Family.

1810. Campanula rapunculoides L. European Bellflower. Cuyahoga, Lorain, Auglaize, Crawford, Carroll, Franklin, Hamilton. From Europe.

1811. Campanula americana L. Tall Bellflower. General and abundant.

1812. Campanula rotundifolia L. Harebell. Ottawa County.

1813. Campanula aparinoides Pursh. Marsh Bellflower. Rather general; no specimens from the southeastern and northwestern counties.

1814. Specularia perfoliata (L.) DC. Venus'-looking-glass. General

Lobeliaceae. Lobelia Family.

1815. Lobelia cardinalis L. Cardinal Lobelia. General.

1816. Lobelia syphalitica L. Blue Lobelia. General and abundant.

1817. Lobelia puberula Mx. Downy Lobelia. Gallia, Meigs, Hocking.

1818. Lobelia spicata Lam. Pale Spiked Lobelia. General.

1819. Lobelia leptostachys A. DC. Spiked Lobelia. Adams, Gallia, Meigs, Hocking, Fairfield, Clark.

1820. Lobelia inflata L. Indian-tobacco. General.

1821. Lobelia kalmii L. Kalm's Lobelia. General.

Order, *Compositales.*

Dipsacaceae. Teazel Family.

1822. Dipsacus sylvestris Huds. Wild Teazel. General and abundant. Naturalized from Europe.

Ambrosiaceae. Ragweed Family.

1823. Xanthium pennsylvanicum Wallr. Pennsylvania Cocklebur. General.

1824. Xanthium americanum Walt. American Cocklebur. Athens, Vinton, Washington.

1825. Xanthium spinosum L. Spiny Cocklebur. Montgomery County. Introduced.

1826. Ambrosia trifida L. Giant Ragweed. General.

1827. Ambrosia psilostachya DC. Western Ragweed. Franklin, Lake. Introduced from the West.

1828. Ambrosia elatior L. Roman Ragweed. General and abundant.

Helianthaceae. Sunflower Family.

1829. Heliopsis helianthoides (L.) Sw. Smooth Oxeye. General and abundant.

1830. Heliopsis scabra Dunal. Rough Oxeye. Erie, Wyandot, Wayne. Madison, Ross.

1831. Verbesina alba L. Verbesina. Rather general.

1832. Rudbeckia triloba L. Thinleaf Cone-flower. Rather general; no specimens from the eastern counties.

1833. Rudbeckia hirta L. Black-eyed-Susan. General and abundant.

1834. Rudbeckia fulgida Ait. Orange Cone-flower. Franklin, Union.

1835. Rudbeckia speciosa Wend. Showy Cone-flower. Montgomery, Champaign, Madison, Franklin.

1835a. Rudbeckia speciosa sullivanti (Boy. & Bead.) Rob. No specimens.

1836. Rudbeckia laciniata L. Tall Cone-flower. General.

1837. Ratibida pinnata (Vent.) Barnh. Tall Nigger-head. Rather general.

1838. Ratibida columnaris (Sims.) D. Don. Prairie Nigger-head. A waif in Franklin County.

1839. Echinacea purpurea (L.) Moench. Purple Cone-flower. Clark, Madison, Franklin, Holmes, Lucas.

1840. Helianthus occidentalis Ridd. Fewleaf Sunflower. Fulton, Erie, Franklin.

1841. Helianthus microcephalus T. & G. Small Wood Sunflower. Rather general.

1842. Helianthus giganteus L. Giant Sunflower. From Erie, Richland, and Fairfield Counties westward.

1843. Helianthus maximiliani Schrad. Maximilian's Sunflower. Lake, Franklin. From the West.

1844. Helianthus grosse-serratus Mart. Sawtooth Sunflower. Cuyahoga, Erie, Huron, Wood, Auglaize, Clark, Madison.

1845. Helianthus kellermani Britt. Kellerman's Sunflower. Franklin County.

1846. Helianthus divaricatus L. Woodland Sunflower. Rather general.

1847. Helianthus mollis. Lam. Hairy Sunflower. Erie, Franklin.

1848. Helianthus doronicoides Lam. Oblong-leaf Sunflower. Rather general; no specimens from the eastern and southeastern counties.

1849. Helianthus decapetalus L. Thinleaf Sunflower. Rather general.

1850. Helianthus tracheliifolius Mill. Throatwort Sunflower. Rather general; no specimens from the southeastern third of Ohio.

1851. Helianthus strumosus L. Paleleaf Wood Sunflower. Rather general; no specimens from the southeastern third of the state.

1852. Helianthus hirsutus Raf. Hirsute Sunflower. General.

1853. Helianthus laetiflorus Pers. Showy Sunflower. Franklin, Wayne.

1854. Helianthus tuberosus L. Jerusalem Artichoke. General.

1855. Helianthus annuus L. Common Sunflower. Rather general. From the West.

1856. Helianthus petiolaris Nutt. Prairie Sunflower. Lake County. From the West.

1857. Phaethusa helianthoides (Mx.) Britt. Sunflower Crownbeard. Madison, Clark, Adams.

1858. Ridan alternifolius (L.) Britt. Ridan. General.

1859. Coreopsis lanceolata L. Lance-leaf Tickseed. Franklin County.

1860. Coreopsis tripteris L. Tall Tickseed. Rather general.

1861. Coreopsis major Walt. Greater Tickseed. Gallia, Lawrence, Scioto.

1862. Coreopsis verticillata L. Whorled Tickseed. No specimens.

1863. Coreopsis tinctoria Nutt. Garden Tickseed. Montgomery, Franklin, Cuyahoga. From the West.

1864. Bidens laevis (L.) B. S. P. Smooth Bur-marigold. Columbiana, Erie, Logan, Hamilton.

1865. Bidens cernua L. Nodding Bur-marigold. General.

1866. Bidens connata Muhl. Swamp Bur-marigold. Rather general.

1867. Bidens comosa (Gr.) Wieg. Leafy-bracted Bur-marigold. Auglaize, Delaware, Franklin, Vinton, Belmont.

1868. Bidens discoidea (T. & G.) Britt. Small Beggar-ticks. Erie County.

1869. Bidens frondosa L. Black Beggar-ticks. Meigs, Vinton, Holmes.

1870. Bidens vulgata Greene. Tall Beggar-ticks. General.

1871. Bidens bipinnata L. Spanish-needles. Rather general; no specimens from the northwestern counties.

1872. Bidens trichosperma (Mx.) Britt. Tall Tickseed. General.

1873. Bidens aristosa (Mx.) Britt. Western Tickseed. Wyandot, Champaign, Clark, Madison.

1874. Megalodonta beckii (Torr.) Greene. Water-marigold. Erie, Stark.

1875. Galinsoga parviflora Cav. Galinsoga. Lake, Cuyahoga, Licking, Belmont, Columbiana, Jefferson, Monroe, Franklin, Montgomery. From tropical America.

1876. Polymnia uvedalia L. Yellow Leaf-cup. Cuyahoga, Noble, Gallia, Lawrence, Clermont, Clark.

1877. Polymnia canadensis L. Small-flowered Leaf-cup. General.

1878. Silphium integrifolium Mx. Entire-leaf Rosin-weed. No specimens.

1879. Silphium trifoliatum L. Whorled Rosin-weed. Rather general; no specimens from the northwestern counties.

1880. Silphium laciniatum L. Compass-plant. Summit County

1881. Silphium perfoliatum L. Indian-cup. General.

1882. Silphium terebinthinaceum Jacq. Prairie Dock (Rosin-weed). Cuyahoga, Wayne, Erie, Ottawa, Lucas, Fulton, Defiance, Hancock, Champaign, Clark, Madison.

1883. Parthenium hysterophorus L. Parthenium. A waif in Franklin County.

1884. Tetraneuris herbacea Greene. Eastern Tetraneuris. Ottawa County.

1885. Helenium autumnale L. Common Sneezeweed. Rather general; no specimens from the southeastern counties.

1886. Helenium nudiflorum Nutt. Purple-headed Sneezeweed. Lake, Franklin.

1887. Helenium tenuifolium Nutt. Slender-leaf Sneezeweed. Franklin, Lake.

1888. Boebera papposa (Vent.) Rydb. Fetid Marigold. Franklin, Delaware, Logan, Madison, Hamilton. From the West.

1889. Inula helenium L. Elecampane. General, but no specimens from the southernmost counties. From Europe.

1890. Gifola germanica (L.) Dum. Herb Impius. Guernsey County. From Europe.

1891. Gnaphalium obtusifolium L. Fragrant Cudweed. General.

1892. Gnaphalium decurrens Ives. Clammy Cudweed. Cuyahoga County.

1893. Gnaphalium uliginosum L. Marsh Cudweed. General.

1894. Gnaphalium purpureum L. Purplish Cudweed. Rather general.

1895. Anaphalis margaritacea (L.) Benth. & Hook. Pearly Everlasting. Cuyahoga County.

1896. Antennaria parlinii Fern. Parlin's Everlasting. General.

1897. Antennaria solitaria Rydb. Single-headed Everlasting. Lawrence County.

1898. Antennaria plantaginifolia (L.) Rich. Plantain-leaf Everlasting. General.

1899. Antennaria neodioica Greene. Smaller Everlasting. Lake, Auglaize.

1900. Antennaria neglecta Greene. Field Everlasting. Rather general.

1901. Grindelia squarrosa (Pursh) Dun. Broadleaf Gum-plant. Hamilton County. From the West.

1902. Chrysopsis graminifolia (Mx.) Ell. Grassleaf Golden-aster. No specimens.

1903. Chrysopsis mariana (L.) Nutt. Maryland Golden-aster. Hocking, Jackson.

1904. Solidago squarrosa Muhl. Stout Goldenrod. Ashtabula, Lake, Cuyahoga.

1905. Solidago caesia L. Wreath Goldenrod. General.

1906. Solidago flexicaulis L. Zig-zag Goldenrod. Eastern Ohio, as far west as Cuyahoga, Fairfield, Jackson, and Lawrence Counties; also in Ottawa County.

1907. Solidago bicolor L. White Goldenrod. Columbiana, Geauga, Cuyahoga, Summit, Wayne, Erie, Fairfield, Vinton, Jackson, Lawrence.

1908. Solidago hispida Muhl. Hairy Goldenrod. Ottawa, Lake.

1909. Solidago erecta Pursh. Slender Goldenrod. Fairfield, Hocking, Meigs.

1910. Solidago uliginosa Nutt. Bog Goldenrod. Lucas, Portage, Stark, Wayne, Licking, Franklin.

1911. Solidago speciosa Nutt. Showy Goldenrod. Lucas, Franklin, Fairfield, Lawrence.

1912. Solidago rigidiuscula (T. & G.) Port. Slender Showy Goldenrod. Erie, Wyandot, Wood, Lucas, Fulton.

1913. Solidago rugosa Mill. Wrinkle-leaf Goldenrod. Rather general.

1914. Solidago patula Muhl. Roughleaf Goldenrod. Rather general.

1915. Solidago ulmnifolia Muhl. Elmleaf Goldenrod. Rather general.

1916. Solidago neglecta T. & G. Swamp Goldenrod. Wood, Madison, Fairfield.

1917. Solidago juncea Ait. Plume Goldenrod. Rather general.

1918. Solidago arguta Ait. Cutleaf Goldenrod. Erie County.

1919. Solidago canadensis L. Canada Goldenrod. General and abundant.

1920. Solidago serotina Ait. Late Goldenrod. General.

1921. Solidago nemoralis Ait. Gray Goldenrod. General.

1922. Solidago rigida L. Stiff Goldenrod. Erie, Ottawa, Lucas, Defiance, Auglaize, Madison, Franklin, Lawrence.

1923. Solidago ohioensis Ridd. Ohio Goldenrod. Stark, Erie, Wyandot, Franklin, Champaign, Clark, Montgomery.

1924. Solidago riddellii Frank. Riddell's Goldenrod. Lucas, Fulton, Wyandot, Franklin, Madison, Clark.

1925. Euthamia graminifolia (L.) Nutt. Bushy Fragrant Goldenrod. General.

1926. Euthamia tenuifolia (Pursh.) Greene. Slender Fragrant Goldenrod. Erie, Lucas, Cuyahoga, Lake.

1927. Bellis perennis L. European Daisy. Lake, Cuyahoga. From Europe.

1928. Boltonia asteroides (L.) L'Her. Boltonia. Erie, Ottawa, Lucas, Auglaize, Paulding, Defiance.

1929. Sericocarpus linifolius (L.) B. S. P. Narrowleaf Whitetop Aster. No specimens.

1930. Sericocarphus asteroides (L.) B. S. P. Toothed Whitetop Aster. Cuyahoga, Summit, Wayne, Holmes, Fairfield, Hocking, Jackson, Gallia, Lawrence.

1931. Aster divaricatus L. White Wood Aster. Meigs, Franklin, Fairfield,. Lorain, Erie.

1932. Aster macrophyllus L. Largeleaf Aster. Rather general; no specimens from the southwestern counties.

1933. Aster shortii Hook. Short's Aster. From Franklin and Montgomery Counties southward; also in Lake and Ottawa Counties.

1934. Aster azureus Lindl. Azure Aster. Franklin, Wood, Fulton, Erie.

1935. Aster cordifolius L. Common Blue Wood Aster. Rather general.

1936. Aster lowrieanus Port. Lowrie's Aster. Lake, Cuyahoga, Auglaize, Fairfield, Hamilton.

1937. Aster lindleyanus T. & G. Lindley's Aster. Wayne, Franklin.

1938. Aster drummondii Lindl. Drummond's Aster. Madison County.

1939. Aster safittifolium Willd. Arrowleaf Aster. Rather general.

1940. Aster undulatus L. Wavy-leaf Aster. Wayne County.

1941. Aster patens Ait. Late Purple Aster. Wayne County.

1942. Aster phlogifolius Muhl. Thinleaf Purple Aster. Wayne, Portage.

1943. Aster novae-angliae L. New England Aster. General.

1944. Aster oblongifolius Nutt. Aromatic Aster. No specimens.

1945. Aster puniceus L. Purple-stem Aster. Rather general.

1946. Aster prenanthoides Muhl. Crooked-stem Aster. Rather general.

1947. Aster laevis L. Smooth Aster. Rather general.

1948. Aster junceus Ait. Rush Aster. Licking, Wayne.

1949. Aster lateriflorus (L.) Britt. Starved Aster. Rather general.

1950. Aster hirsuticaulis Lindl. Roughstem Aster. Warren, Auglaize.

1951. Aster vimineus Lam. Small White Aster. Wayne County.

1952. Aster multiflorus Ait. Dense-flowered Aster. Lucas, Erie, Gallia.

1953. Aster dumosus L. Bushy Aster. Erie County.

1954. Aster salicifolius Lam. Willow Aster. Wayne County.

1955. Aster paniculatus Lam. Panicled Aster. General.

1956. Aster tradescanti L. Tradescant's Aster. Rather general.

1957. Aster faxoni Porter. Faxon's Aster. Vinton County.

1958. Aster ericoides L. White Heath Aster. General.

1958a. Aster ericoides platyphyllus T. & G. Western half of state, east to Erie, Franklin, and Meigs Counties.

1959. Aster ptarmicoides (Nees.) T. & G. Upland White Aster. (Ottawa County—Moseley Herbarium.)

1960. Erigeron pulchellus Mx. Showy Fleabane. General.

1961. Erigeron philadelphicus L. Philadelphia Fleabane. General.

1962. Erigeron annuus (L.) Pers. White-top Fleabane. General.

1963. Erigeron ramosus (Walt.) B. S. P. Daisy Fleabane. General.

1964. Leptilon canadense (L.) Britt. Common Horseweed. General.

1965. Doellingeria umbellata (Mill.) Nees. Tall White-top Aster. Rather general.

1966. Doellingeria infirma (Mx.) Greene. Infirm Aster. Portage County.

1967. Ionactis linariifolius (L.) Greene. Stiffleaf Aster. Adams, Hocking.

1968. Eupatorium maculatum L. Spooted Joe-Pye-weed. General.

1969. Eupatorium purpureum L. Joe-Pye-weed. General.

1970. Eupatorium serotinum Mx. Late-flowering Thoroughwort. Hamilton County.

1971. Eupatorium altissimum L. Tall Thoroughwort. Hamilton, Montgomery, Franklin, Erie, Lucas.

1972. Eupatorium sessilifolium L. Upland Boneset. Southeastern half of state, to Montgomery, Franklin, Wayne, Portage and Hamilton Counties.

1973. Eupatorium rotundifolium L. Roundleaf Thoroughwort. Hocking County.

1974. Eupatorium perfoliatum L. Common Boneset. General.

1975. Eupatorium urticaefolium Reich. White Snake-root. General.

1976. Eupatorium aromaticum L. Smaller White Snake-root. Hocking County.

1977. Eupatorium coelestinum L. Mist-flower. Southern Ohio; north to Hamilton, Fairfield, and Washington Counties; also in Ashtabula County.

1978. Kuhnia eupatorioides L. False Boneset. Lucas, Erie, Clark, Franklin, Gallia, Lawrence.

1979. Lacinaria squarrosa (L.) Hill. Scaly Blazing-star. Lucas, Erie.

1980. Lacinaria cylindrica (Mx.) Ktz. Cylindric Blazing-star. Franklin County.

1981. Lacinaria punctata (Hook.) Ktz. Dotted Blazing-star. A waif in Franklin County.

1982. Lacinaria scariosa (L.) Hill. Large Blazing-star. Erie, Lucas, Fairfield.

1983. Lacinaria spicata (L.) Ktz. Dense Blazing-star. Paulding, Lucas, Wood, Erie, Wyandot, Champaign, Clark, Hocking.

1984. Vernonia noveboracensis (L.) Willd. New York Ironweed. Gallia County.

1985. Vernonia altissima Nutt. (V. maxima Small). Tall Ironweed. General and abundant.

1986. Vernonia fasciculata Mx. Western Ironweed. Erie County.

1987. Vernonia missurica Raf. Missouri Ironweed. Erie County.

1988. Elephantopus carolinianus Willd. Carolina Eilephant's-foot. Scioto, Jackson.

1989. Achillea millefolium L. Common Milfoil. General and abundant.

1990. Anthemis cotula L. Common Dog-fennel. General and abundant. Naturalized from Europe.

1991. Anthemis arvensis L. Field Dog-fennel. Lorain, Lake. From Europe.

1992. Anthemis tinctoria L. Yellow Dog-fennel. Guernsey County. From Europe.

1993. Chrysanthemum leucanthemum L. Oxeye Daisy. General and abundant. Naturalized from Europe.

1994. Chrysanthemum parthenium (L.) Pers. Common Feverfew. Lake, Erie, Montgomery. From Europe.

1995. Chrysanthemum balsamita L. Sweet-Mary. Cuyahoga, Ottawa, Franklin, Madison. Escaped from gardens.

1996. Chrysanthemum indicum L. Chrysanthemum. Escaped in Adams County.

1997. Matricaria inodora L. Scentless Camomile. Lake, Lawrence. From Europe.

1998. Matricaria chamomilla L. German Camomile. Ottawa County. From Europe.

1999. Matricaria matricarioides (Lees.) Port. Rayless Camomile. Stark County. From the Pacific coast.

2000. Tanacetum vulgare L. Common Tansy. General. Naturalized from Europe.

2001. Artemisia caudata Mx. Wild Wormwood. Erie County.

2002. Artemisia annua L. Annual Wormwood. Rather general. Introduced.

2003. Artemisia biennis Willd. Biennial Wormwood. Ashtabula, Cuyahoga, Lucas, Auglaize, Shelby, Wyandot, Franklin.

2004. Artemisia vulgaris L. Common Mugwort. Escaped in Lake and Cuyahoga Counties.

2005 Artemisia pontica L. Roman Wormwood. Champaign, Portage. From Europe.

2006. Artemisia gnaphalodes Nutt. Prairie Cudweed. Lake County. From the West.

2007 Erechtites hieracifolia (L.) Raf. Fireweed. General.

2008. Mesadenia reniformis (Muhl.) Raf. Great Indian-plantain. Clermont, Greene.

2009. Mesadenia atriplicifolia (L.) Raf. Pale Indian-plantain. General.

2010. Mesadenia tuberosa (Nutt.) Britt. Tuberous Indian-plantain. Montgomery, Champaign, Logan.

2011. Synosma suaveolens (L.) Raf. Sweet-scented Indian-plantain. Cuyahoga, Lorain, Stark, Clark, Jackson.

2012. Senecio aureus L. Golden Squaw-weed. General.

2013. Senecio obvatus Muhl. Roundleaf Squaw-weed. Rather general.

2014. Senecio panperculus Mx. Balsam Squaw-weed. Ottawa County.

2015. Senecio vulgaris L. Common Groundsel. Lake, Lorain, Auglaize. From Europe.

2016. Tussilago farfara L. Coltsfoot. Lake, Cuyahoga. From Europe.

2017. Arctium tomentosum (Lam.) Schk. Woolly Burdock. Erie County. From Europe.

2018. Arctium lappa L. Great Burdock. Lorain, Cuyahoga. From Europe.

2019. Arctium minus Schk. Common Burdock. General and abundant. Naturalized from Europe.

2020. Cirsium lanceolatum (L.) Hill. Spear Thistle. Rather general. From Europe.

2021. Cirsium altissimum (L.) Spreng. Tall Thistle. Rather general.

2022. Cirsium discolor (Muhl.) Spreng. Field Thistle. Western Ohio, as far east as Erie, Huron, Fairfield, and Clermont Counties.

2023. Cirsium virginianum (L.) Mx. Virginia Thistle. Madison County.

2025. Cirsium muticum Mx. Swamp Thistle. General.

2026. Cirsium arvense (L.) Scop. Canada Thistle. General. From Europe.

2027. Onopordon acanthium L. Scotch Thistle. Hamilton, Wayne. From Europe.

2028. Centaurea scabiosa L. Scabious Star-thistle. Lake County. From Europe.

2029. Centaurea jacea L. Brown Star-thistle. Richland County. From Europe.

2030. Centaurea cyanus L. Bachelor's-Button. Montgomery, Franklin, Sandusky. Escaped from gardens.

Cichoriaceae. Chicory Family.

2031. Cichorium intybus L. Chicory. Rather general. Introduced from Europe.

2032. Cynthia virginica (L.) Don. Virginia Cynthia. General.

2033. Lapsana communis L. Nipplewort. Franklin, Lake. From Europe.

2034. Arnoseris minima (L.) Dum. Lamb Succory. Lake County. From Europe.

2035. Hypochaeris radicata L. Long-rooted Cat's-ear. Lake, Ashtabula. From Europe.

2036. Apargia nudicaulis (L.) Britt. Rough Hawkbit. Lake County. From Europe.

2037. Tragopogon pratensis L. Yellow Goat's-beard. Lake, Erie, Fulton, Auglaize, Franklin, Miami. From Europe.

2038. Tragopogon porrifolius L. Salsify. Rather general; no specimens from the southeastern third of the state. From Europe.

2039 Sonchus arvensis L. Field Sow-thistle. Lake, Franklin. From Europe.

2040. Sonchus oleraceus L. Common Sow-thistle. General. Naturalized from Europe.

2041. Sonchus asper (L.) Hill. Spiny Sow-thistle. General. Naturalized from Europe.

2042. Lactuca virosa L. Prickly Lettuce. General and abundant. Naturalized from Europe.

2043. Lactuca saligna L. Willow Lettuce. Franklin, Greene, Montgomery. From Europe.

2044. Lactuca hirsuta Muhl. Hairy Lettuce. Tuscarawas, Ross, Union.

2045. Lactuca canadensis L. Tall Lettuce. General.

2046. Lactuca sagittifolia Ell. Arrowleaf Lettuce. Fairfield County.

2047 Lactuca villosa Jacq. Hairy-veined Blue Lettuce. Miami, Montgomery.

2048. Lactuca floridana (L.) Gaertn. Florida Lettuce. From Erie, Franklin, and Ross Counties westward.

2049 Lactuca spicata (Lam.) Hitch. Tall Blue Lettuce. Rather general.

2049a. Lactuca spicata aurea Jennings. Holmes, Defiance, Cuyahoga, Franklin, Athens.

2050. Nabalus altissimus (L.) Hook. Tall Rattlesnake-root. General.

2051. Nabalus albus (L.) Hook. White Rattlesnake-root. General.

2052. Nabalus asper (Mx.) T. & G. Rough Rattlesnake-root. Erie County.

2053. Nabalus racemosus (Mx.) DC. Glaucous Rattlesnake-root. Fulton, Lucas, Ottawa, Erie, Huron, Wyandot, Champaign, Clark.

2054. Nabalus crepidineus (Mx.) DC. Corymbed Rattlesnake-root. Cuyahoga, Champaign, Warren.

2055. Hieracium canadense Mx. Canada Hawkweed. Erie, Cuyahoga.

2056. Hieracium paniculatum L. Panicled Hawkweed. Cuyahoga, Wayne, Richland, Fairfield, Monroe

2057 Hieracium scabrum Mx. Rough Hawkweed. General.

2058. Hieracium gronovii L. Gronovius' Hawkweed. Fulton, Erie, Franklin, Gallia.

2059. Hieracium marianum Willd. Maryland Hawkweed. No specimens.

2060. Hieracium venosum L. Veined Hawkweed. Eastern Ohio; west to Cuyahoga, Knox, Fairfield, Jackson, and Lawrence Counties.

2061. Hieracium greenii Port. & Britt. Green's Hawkweed. No specimens.

2062. Hieracium pilosella L. Mouse-ear Hawkweed. Lake County. From Europe.

2063. Hieracium aurantiacum L. Orange Hawkweed. Ashtabula, Geauga. From Europe.

2064. Crepis capillaris (L.) Wallr. Smooth Hawksbeard. Lake County. From Europe.

2065. Leontodon taraxacum L. Dandelion. General and very abundant. Naturalized from Europe.

INDEX TO THE GENERA

The numbers refer to the list numbers at the left of the species names. A few familiar synonyms have been included.

Bulletins Ohio Biological Survey

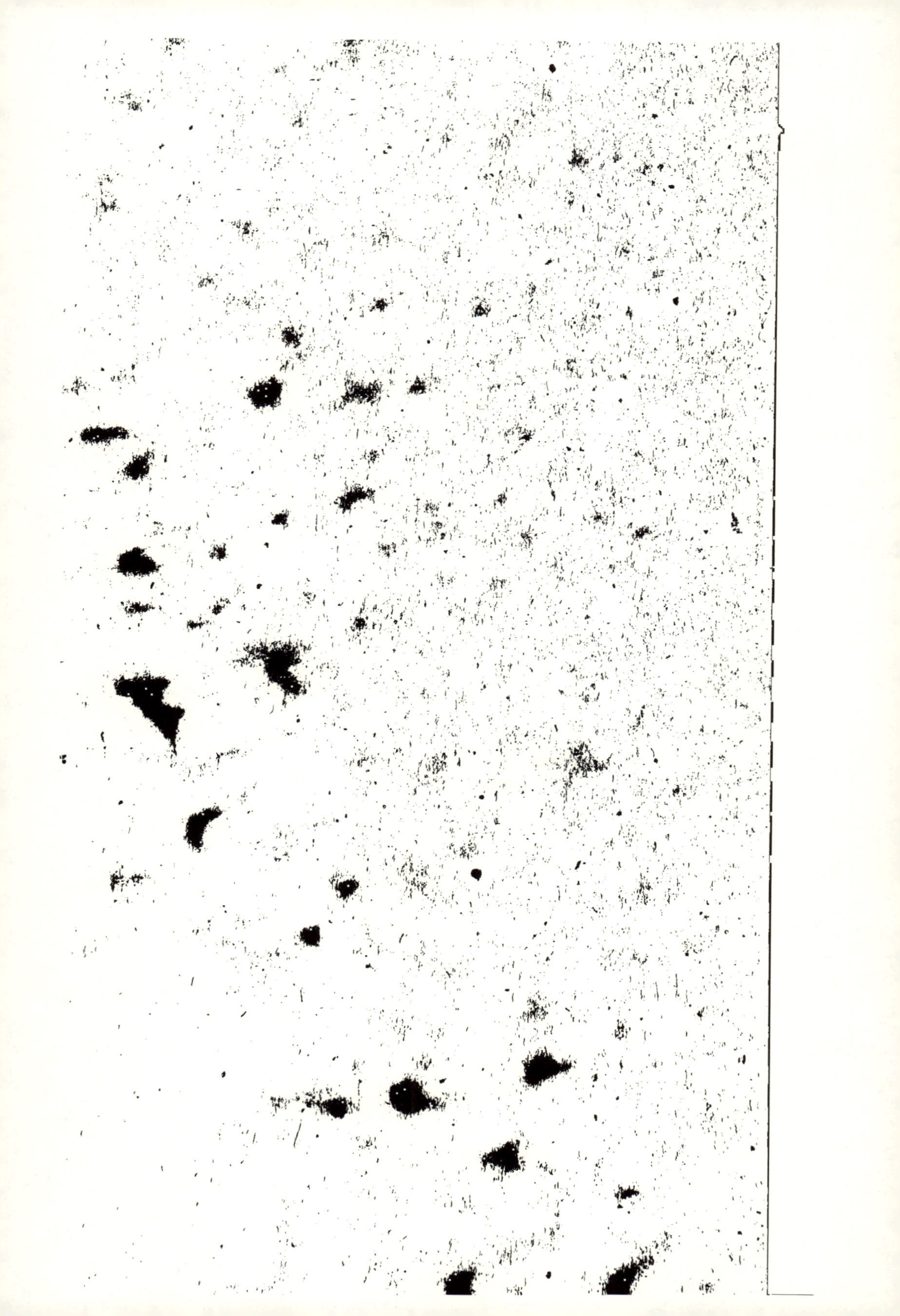

Made in the USA
Columbia, SC
31 October 2023